Reverse Engineering Code with IDA Pro

Justin Ferguson
Dan Kaminsky
Jason Larsen
Luis Miras
Walter Pearce

KEY	SERIAL NUMBER
001	HJIRTCV764
002	PO9873D5FG
003	829KM8NJH2
004	BAL923457U
005	CVPLQ6WQ23
006	VBP965T5T5
007	HJJJ863WD3E
008	2987GVTWMK
009	629MP5SDJT
010	IMWQ295T6T

PUBLISHED BY
Syngress Publishing, Inc.
Elsevier, Inc.
30 Corporate Drive
Burlington, MA 01803

Reverse Engineering Code with IDA Pro

Printed in the United States of America
1 2 3 4 5 6 7 8 9 0

ISBN 13: 978-1-59749-237-9

Publisher: Andrew Williams
Technical Editor: Dan Kaminsky
Project Manager: Anne McGee
Page Layout and Art: SPi

For information on rights, translations, and bulk sales, contact Matt Pedersen, Commercial Sales Director and Rights, at Syngress Publishing; email m.pedersen@elsevier.com.

About IOActive

Established in 1998, IOActive has successfully positioned itself as an industry leader in the Northwest's computer security community, where it specializes in infrastructure assessment services, application security services, managed services, incident response services, and education services. The company has helped various Fortune 500 organizations with services ranging from enterprise risk management to independent technical validations of security hardware and a wide range of applications. It has also been commissioned to work on IT disaster recovery and business continuity planning for major insurance companies, state organizations and energy companies. IOActive's consultants are members and active contributors to local and nationally recognized computer security organizations such as SANS, Agora, CRIME, ISSA, CTIN, WSA, HoneyNet Research Alliance, OWASP, and the University of Washington Information Assurance School.

Technical Editor and Contributing Author

Dan Kaminsky is the Director of Penetration Testing for IOActive. Previously of Cisco and Avaya, Dan has been operating professionally in the security space since 1999. He is best known for his "Black Ops" series of talks at the well respected Black Hat Briefings conferences. He is also the only speaker who has attended and spoken at every single "Blue Hat" Microsoft internal training event. Dan focuses on design level fault analysis, particularly against massive-scale network applications. Dan regularly collects detailed data on the health of the worldwide Internet, and recently used this data to detect the worldwide proliferation of a major rootkit. Dan is one of the few individuals in the world to combine both technical expertise with executive level consulting skills and prowess.

Contributing Authors

Justin Ferguson is a security consultant and researcher at IOActive. He is involved with helping Fortune 500 companies understand and mitigate risk introduced in complex software computing environments via the Application Security Practice at IOActive. Justin has over six years experience working as a reverse engineer, source code auditor, malware analyst, and enterprise security analyst for industries ranging from financial institutions to the federal government.

I would like to thank my father, Bruce Dennis Ferguson, who was a great man; I regret never having apologized to you nor allowing you to see the man your son has become. I would like to thank all of the blue collar union workers from Boston who worked themselves to the bone to make sure their children had a better life. No mention of these men would be complete if I neglected the women who stood by their sides and saw them through each day; you all truly are beautiful. I'd like to take a moment to remember everyone from the South End and Brockton/South Shore who didn't make it and for those still struggling; continue on with the belief that unearned suffering is redemptive. Saint Jude, pray for us all.

Jason Larsen has penetrated and owned some of the most integral systems on the planet. His career began when he was at Idaho State University and detected Internet-wide stealth scanning. He was awarded two scholarships in order to support his research into and creation of detection systems, including authorship of one of the first Intrusion Prevention Systems that actually blocked penetration. Mr. Larsen has been unable to publish most of his work due to national security concerns. His work for the Department of Energy through the Idaho National Laboratories allowed him to develop even more elegant solutions to the security problems of major SCADA and PCS systems. His security work has benefited hundreds of clients among several industries, including US and foreign.

I'd like to dedicate this book to the infinite patience and understanding of The Girlfriend. Thank you for the quiet nods when listening to the latest problem and the occasional push out the door to get some sunlight. Every geek should be required to have a permanent tattooed companion.

Luis Miras is an independent security researcher. He has worked for both security product vendors and leading consulting firms. His interests include vulnerability research, binary analysis, and hardware/software reversal. In the past, he has worked in digital design and embedded programming. He has presented at CanSecWest, Black Hat, CCC Congress, XCon, REcon, DefCon, and other conferences worldwide. When he isn't heads down in IDA or a circuit board, you will likely find him boarding down some sweet powder.

I dedicate this book to my parents and brothers. I would like to thank Don Omar, Sister Nancy, and Nas for providing the coding soundtrack. I would like to send greetz to all my friends and let them know that, yes, I'm alive and no longer MIA. Thanks to Sebastian "topo" Muniz for the IDA discussions and bouncing ideas.

Walter Pearce provides application security and penetration testing services for IOActive, and is a regular contributor to the ongoing research and development of advanced tools that automate IT security testing and protective functions. His career began at 12, and his first professional role was as the operator of a data center cluster for an online retailer, which led to Senior Programming Engineer positions at financial service firms and institutions. During his time in the finance industry, Walter specialized in the conception of internal threats and designed mitigations to reduce incidence of such events. Mr. Pearce is often requested by clients to provide expert application security services involving a variety of platforms and languages.

To Becca, Mom, David. Love ya all.

Contents

Chapter 1

Introduction

The theater of the information security professional has changed drastically in recent years. We are no longer tasked with defending critical organizational assets from the unwelcome inquiry of curious youth; we, as a community, are now faced with fending off relentless and technically sophisticated attacks perpetrated by organized and nation state-backed criminals motivated by financial or geopolitical gain.

The prevalence of security holes in programs and protocols, the increasing size and complexity of the Internet, and the sensitivity of the information stored throughout have created a target-rich environment for our next-generation adversary. This criminal element is employing advanced polymorphic software that is specifically engineered to evade IDS, IPS and AV detection engines, and provide complete remote control and eavesdropping functionality on the victims' computer. One of the few offenses we can deploy in order to understand and predict the impact of these malicious software programs is through employment of advanced reverse engineering techniques, leveraging industry-standard tools from companies like Data Rescue and Zynamics.

This book represents the leading thought from the reverse engineering world. The authors are tremendous people in their own right, and I trust you and your organization will find a wealth of information that will help prepare you for the proactive computer security frontier.

A big thanks to Lauren Vogt, Ted Ipsen, Dan Kaminsky, Jason Larsen, Walter Pearce, Justin Ferguson, Luis Miras, and the kind folks at Syngress for making this book possible.

Joshua J. Pennell, Founder and CEO
IOActive, Inc. Comprehensive Computer Security Services

An Overview of Code Debuggers

Sooner or later you will want to know absolutely everything about an executable file. You may want to know, for instance:

- The exact memory address that it is calling
- The exact region of memory that it is writing to
- What region it's reading from
- Which registers it's making use of

Debuggers will aid you in reverse-engineering a file for which you don't have the source code, by disassembling the file in question. This comes in handy when you're analyzing malware, as you almost never have access to the executable's original source code. The goal of this section is not to coach you in depth on how to use these debuggers, but simply to show you that they are out there and available for you to use. Debuggers are very powerful tools that take a long time to learn to use to their fullest extent.

The "cream of the crop" in debuggers and the focus of this book is Interactive Disassembler Pro (IDA Pro), available from DataRescue. IDA Pro should be your first choice of debuggers for an enterprise environment. It isn't really expensive, and is well worth the nominal outlay for the features it offers.

TIP

DataRescue offers a demo version from their Web site at www.datarescue.com/idabase/index.htm. This version can only work with a limited range of file and processor types, is time limited, runs only as a Windows GUI application, and so on.

IDA Pro is much more than a simple debugger. It is a programmable, interactive disassembler and debugger. With IDA Pro you can reverse-engineer just about any type of executable or application file in existence. IDA Pro can handle files from console machines such as Xbox, Playstation, Nintendo, to Macintosh computer systems, to PDA platforms, Windows, UNIX, and a whole lot more. Figure 1.1 shows the initial load screen wizard when you first start IDA Pro. Notice all the file types and tabs that will help you select the proper analysis for the file type that you wish to disassemble.

Figure 1.1 IDA Pro's Disassembly Database Chooser Loads Upon Start

In Figure 1.2, IDA Pro has loaded and is disassembling a WootBot variant with file name *instantmsgrs.exe*. Part of what we can see from Figure 1.2 is that *instanmsgrs.exe* was packed using an executable packer called Molebox. You can also plainly see the memory calls that it's making, and the Windows DLLs that are being called. This type of information can be invaluable when it comes to fighting off a virus or malware outbreak, especially if you need to make a custom cleaner in order to repair your systems.

Figure 1.2 IDA Pro Disassembles *instantmsgrs.exe*, a WootBot Variant

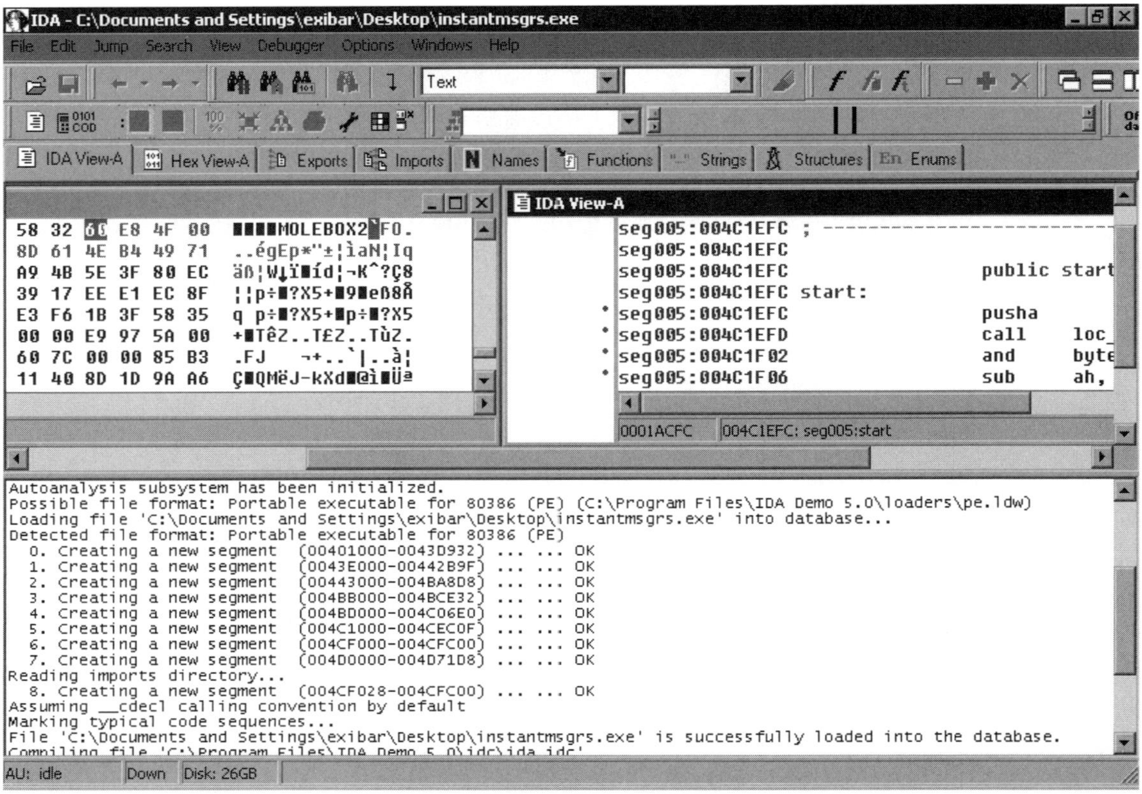

Summary

IDA is one of the most popular debugging tools for Windows. First, IDA Pro is a disassembler, in that it shows the assembly code of a binary (an executable or a dynamic link library [DLL]). It also comes with advanced features that try to make understanding the assembly code as easy as possible. Second, it is also a debugger, in that it allows the user to step through the binary file to determine the actual instructions being executed, and the sequence in which the execution occurs. You'll learn about all of these features throughout this book. IDA Pro is widely used for malware analysis and software vulnerability research, among other purposes. IDA Pro can be purchased at www.datarescue.com.

Chapter 2

Assembly and Reverse Engineering Basics

Solutions in this chapter:

- **Assembly and the IA-32 Processor**
- **The Stack, the Heap and Other Sections of a Binary Executable**
- **IA-32 Instruction Set Refresher and Reference**

☑ **Summary**

Introduction

In this chapter we will introduce basic items, providing a brief introduction to assembly and the Intel architecture processor and covering various other concepts that will help ease you into the subject matter. This book focuses on 32-bit Intel architecture (IA-32) assembly and deals with both the Windows and Linux operating systems. The reader is expected to be at least mildly familiar with IA-32 assembly (although the architecture and instruction set are covered to some degree in this chapter), and a firm grasp of C/C++ is expected. The point of this chapter is simply to give those who are either unfamiliar with or a little rusty on the subjects presented in this book a base to work from, and to provide a basic reference point to which the authors can refer should it be deemed necessary.

Assembly and the IA-32 Processor

Assembly is an interesting method for communicating with computers; Donald Knuth once said that "Science is what we understand well enough to explain to a computer. Art is everything else we do." To me, this truth is most prevalent in assembly programming, because in so many areas as you write assembly you find yourself doing things like abusing instructions by using them for something other than their intended purpose (for instance, using the load effective address (LEA) instruction to do something other than pointer arithmetic). But what is assembly exactly? Assembly refers to the use of instruction mnemonics that have a direct one-to-one mapping with the processor's instruction set; that is to say, there are no more layers of abstraction between your code and the processor: what you write is what it gets (although there are some exceptions to this on some platforms where the assembler exports pseudo-instructions and translates them into multiple processor instructions—but the prior statement is generally true).

If we take, for instance, the following single line of C code:

```
return 0;
```

we may end up with the following assembly being generated:

```
leave
xor    eax,    eax
ret
```

NOTE

Many assemblers have different syntaxes. In the assembly code above, and indeed in assembly used throughout most of this book, the syntax employed is Intel syntax. Another popular syntax used largely in the Unix world is AT&T syntax, which looks a little bit different. The choice of Intel was made because it is the syntax used by IDA in its disassemblies and is indeed a more popular syntax overall. This means that there will generally be more books, white papers and people willing to answer questions when you use the Intel syntax.

The difference between AT&T and Intel syntax is outside the scope of this document, but just so you know what it looks like, the following example is given:

Intel:

```
  leave
xor      eax, eax
ret
```

AT&T:

```
  leave
xorl     %eax, %eax
ret
```

Note, however, that AT&T is still in use in many places and is generally the standard syntax used in the Unix world (although this is slowly changing on IA-32-based computers running Unix and Unix-like OSes). Therefore, it may not be a bad idea to spend a few extra clock cycles at least learning it, especially if you intend to do much work on the various Unices or other platforms.

Don't worry if at the moment you're not entirely sure what that means; I'm just hoping to get you into the groove of assembly. Just know that the previous code is IA-32 assembly, and it is the same as saying "return 0" in C. But this isn't what the processor sees; assembly is the last layer of code that is considered human-readable. When your compiler comes through and compiles and assembles the code, it outputs what are known as "opcodes," which are the binary representations of the instructions, or the on-off sequences necessary to execute individual instructions. Opcodes are typically represented in hexadecimal for humans though, since it tends to be easier to read than binary. The opcodes for the previous instructions are as follows:

```
0xC9            (leave)
0x31, 0xc0      (xor eax, eax)
0xC9            (ret)
```

As we see, these are the three basic layers of abstraction and, as advances in computing continue, we add more and more layers of abstraction, such as virtual machines used by Java and .NET applications. However, *everything* in the end is assembly, and that is just fixed sequences of ones and zeros being sent to the processor. For a more complete discussion of opcodes please refer to the *Intel® 64 and IA-32 Architectures Software Developer's Manual Volume 2A*, section 2.1.2.

Tools & Traps...

Opcodes and shellcode

In almost everyone's transition into the digitally sublime, we encounter the term *shellcode*, and it strikes fear deep into our hearts. We see these character arrays of cryptic hexadecimal numbers and we're just not quite sure what they do. Anyone who has examined an exploit has probably run across it, and if you're early enough into your career you probably don't fully understand it.

Rest assured, however, that these mystical thoughts about it are overcomplicated. Shellcode is simply a series of opcodes, typically stored in a C character array. The term *shellcode* derives from the fact that the series of opcodes are the instructions necessary to execute a shell, such as /bin/sh or cmd.exe. In our case, if we take our previous example:

```
return 0;
```

to generate shellcode for that particular C level instruction, we would simply use the opcodes for that instruction, which are 0xC9, 0x31, 0xC0, 0xC9; if we put it into a C program it would probably look like this:

```
unsigned char shellcode[] = "\xc9\x31\xc0\xc9";
```

Now that you know that, you might be inclined to feel a little silly for thinking it much more complicated than it is, but you shouldn't. I think everyone goes through that stage in their path towards enlightenment—I know I did.

So now we have some comprehension of what assembly instructions are, but how are they used? An instruction that takes an argument (also known as an operand) will, depending on the instruction, either take a constant, a memory variable or a register. Constants are simple; they are statically defined in the source code. For instance, if a section of code were to use the following instruction:

```
mov eax, 0x1234
```

then the hexadecimal number 0x1234 would be the constant. Constants are fairly straightforward and there is not much to say about them aside from the fact that they're constant and are typically encoded directly into the instruction. One interesting subject, though, is that if you consider our prior example of returning zero in C, the astute reader may note that the assembly that was generated by the compiler doesn't contain a constant even though the source-level code does contain one. This is the result of an optimization performed by the compiler and its recognizing that copying zero is a larger instruction than performing an exclusive or.

Next, we encounter registers. Registers are somewhat akin to a variable in C/C++. The general purpose register can contain an integer, an offset, an immediate value, a pointer or really anything that can be represented in 32 bits. It is basically a preallocated variable that physically exists in the processor and is always in scope. They are used a little differently from what we typically think of variables, however, as they are used and reused over and over again, whereas in a C or C++ program we will usually define a variable for a single purpose and then never use it again for something else.

In IA-32 there are eight 32-bit general purpose registers, six 16-bit segment registers, one 32-bit instruction pointer register, one 32-bit status register, five control registers, three memory management registers, eight debug registers, and so on. In most cases you will only be dealing with the general purpose registers, the instruction pointer, the segment registers and the status register. If you're dealing with OS drivers or similar, you're more likely to encounter the other registers. Here we're going to cover the general purpose registers, the instruction pointer, the status register and the segment registers. As for the others, it's probably good enough just to know they exist, although naturally the interested reader is encouraged to consult the Intel documentation for further details.

The eight 32-bit general purpose registers are as follows: EAX, EBX, ECX, EDX, ESI, EDI, EBP and ESP. These registers can mostly be used as one sees fit, with a few notable exceptions. For instance, many instructions assign specific registers to certain arguments (or operands). As a specific example, many of the string instructions use ECX as a counter, ESI as a source pointer and EDI as a destination pointer. Furthermore, some of these instructions imply the use of a given segment as a base address in certain memory models (both are covered shortly). Finally, other registers are implied during certain operations. For instance, the EBP and ESP registers are used in many stack operations and their values containing an address that is not mapped into the current process's address space will often result in the application crashing. The IA-32 architecture is almost entirely backwards compatible to the 8086 processor and this is reflected in their registers; all of the general purpose registers can be accessed in a manner yielding the register's full 32-bit contents or its lower 16 bits, and the EAX, EBX, ECX and EDX registers can have their high-order and lower-order 8 bits accessed as well. This is accomplished by using the names reflected in Figure 2.1. For instance, to access the low-order 8 bits of the EAX register, you would replace EAX in your instruction with AL; to access the lower 16 bits of the EBP register you would replace EBP with BP; and to access the second set of 8 bits in the low-order 16 bits of the EDX register, you would replace EDX with DH. In addition to the general purpose registers, we also have the instruction pointer, EIP. The EIP register points to the next instruction to be executed by the processor and by implication the goal of almost any application-based attack is to control this register. Unlike the general purpose registers, however, it cannot be directly modified. This is to say that you cannot execute an instruction to move a value into the register, but rather you would have to perform a set of operations that indirectly modify its value, such as a push onto the stack segment followed by a *ret* instruction. Don't worry if that last sentence was a bit outlandish and you didn't quite understand it yet; both of the instructions referenced and the stack segment will be covered in detail later on in the chapter. Just for now, know that you cannot directly modify the value of the instruction pointer.

Figure 2.1 General Purpose Registers

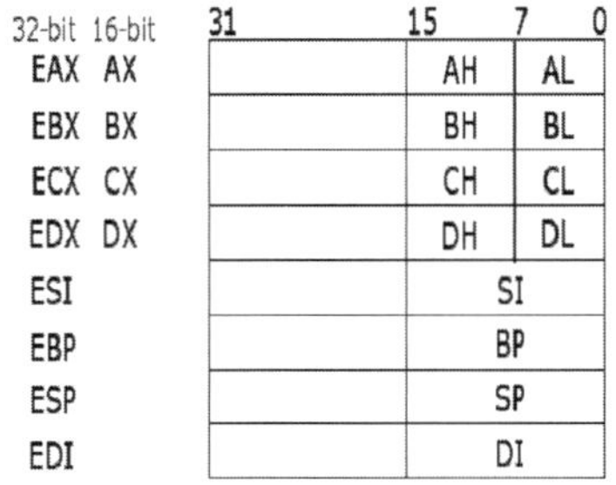

In addition to the EIP register and the general purpose register, we have six 16-bit segment registers: code segment (CS), data segment (DS), stack segment (SS), extra segment (ES), and FS and GS, which are extra general purpose segments. The segment registers contain a pointer to what are called segment selectors and these are often used as a base address from which to take an offset; for instance, consider the following instruction:

```
mov DS:[eax], ebx
```

In this instruction, the contents of the EBX register are copied into an offset into the data segment specified by EAX. Think of this as saying "the address of DS plus the contents of EAX." Segment selectors are a 16-bit identifier for segments, which is to say the segment selector does not point directly to the segment, but rather to a segment descriptor that defines the segment. So, segment registers point to segment selectors, which are used to identify one of the 8192 possible segment descriptors that identify segments. Confused yet?

The segment selectors are relatively simple structures. The bits at 3 through 15 are used as an index into one of the descriptor tables (one of the three memory management registers), bit 2 specifies which descriptor table exactly, and finally the low-order two bits specify a requested privilege level (ranging 0 through 3—privilege levels are discussed later in the chapter). Segment descriptors, while interesting and fairly important to OS design, are again not covered in order to ensure only information relevant to your general purposes is contained in this chapter. As always, the interested reader is encouraged to refer back to the

Intel developer manuals. Finally, in the description of the various relevant registers, we have the EFLAGS register, which contains groups of various flags that indicate the states of previous instructions, status, and things such as the current privilege level and whether interrupts are enabled or not. In the current context of things, the EFLAGS register can't really make sense until we have a better grasp of some of the instructions that make use of it, and as such a thorough description of it is reserved until later in the chapter.

Now that we've described both constant and register operands, it's time to discuss memory operands. Memory operands can be a little bit tricky, although at this stage in the game their description is pretty limited. A memory operand is by and large what a high-level language programmer thinks of as a variable. That is to say that when you declare a variable in a language like C or C++, it's by and large going to exist in memory and thus it will often be a memory operand. These are typically accessed through a pointer, which when dereferenced will either result in the value being loaded into a register or accessed directly from memory. The concept itself is pretty simple, but truly understanding what is going on requires a bit deeper knowledge of how memory is addressed, which in turn depends on the memory model, mode of operation and privilege level being used. This provides an excellent lead into the next few paragraphs, which cover mode of operation.

In IA-32, there are three basic operating modes and one pseudomode of operation. They are *protected mode*, *real-address mode* and *system management mode*, with the pseudomode being a subset of protected mode called *virtual-8086 mode*. In the interest of brevity, only protected mode will be discussed in detail. The biggest difference between the various operating modes is that they modify what instructions and architectural features are present. For instance, RM (also often called *real mode*) is meant for backwards compatibility, and in RM only the real-address mode memory model is supported. The main thing of importance to note here (unless you're reversing old DOS applications or similar), is that when you reset or first power up an IA-32 PC, it is in real mode natively. System management mode (SMM), which has been in the Intel architecture since the 80386 and is used in implementing things such as power management and system hardware control, is also used. It basically stops all other operations and switches to a new address space. Generally speaking, however, almost everything you encounter and use will be in protected mode.

Protected mode represented a huge advancement in the Intel architecture when it was introduced on the 80286 processor and further refined on the 80386 processor. One key issue was that previous versions of the processor supported only one mode of operation and had no inherent hardware-enforced protections on instructions and memory. This not only allowed rogue operators to do anything they wanted, but it also allowed for a faulty application to crash the entire system, so it was an issue of both reliability and security. Another issue with prior processor versions was the 640KB barrier; this is something else that PM overcame. Furthermore, there were other advances, such as hardware-supported multitasking and modifications in how interrupts were handled. Indeed, the 286 and 386 represented significant advances in personal computing.

In the earlier days, such as with the 8086/80186 processor, or today when a modern processor is in real-address mode, the segment register represents the high-order 16 bits of a linear address, whereas in protected mode the selector is an index into one of the descriptor tables. Furthermore, as previously mentioned, with earlier CPUs there was no protection of memory or limit to instructions; in protected mode, there are four privilege levels called *rings*. These rings are given numbers for identification ranging from 0 to 3, with the lowest number being the highest privilege level. Ring-0 is typically used for the operating system, whereas ring-3 is where applications typically run. This protects the OS's data structures and objects from modification by a broken or rogue application and restricts instructions that those applications can run (what good would the levels be if the ring-3 application could just switch its privilege level?). In IA-32 there are three places to find privilege level: in the low-order 2 bits of the CS register there is the Current Privilege Level (CPL), in the low-order 2 bits of a segment descriptor is the Descriptor Privilege Level (DPL), and in the low-order 2 bits of a segment selector is the Requestor's Privilege Level (RPL). The CPL is the privilege level of the currently executing code, the DPL is the privilege level of the given descriptor, and, somewhat obviously, the RPL should be the privilege level of the code that created that segment.

The privilege levels restrict access to more trusted components of the system's data; for instance, a ring-3 application cannot access the data of a ring-2 component. However, the ring-2 component can access the data of a ring-3 component. This is why you can't arbitrarily read data from the Windows or Linux kernel but it can read yours. Another function of this privilege-level separation is that it performs checks on execution transfer control. A request to change execution from the current segment to another causes a check to be performed, ensuring that the CPL is the same as the segment's DPL. Indirect transfers of execution occur via things such as call gates (which are briefly covered later). Finally, privilege levels restrict access to certain instructions that would fundamentally change the environment of the OS or operations that are typically reserved for the OS (such as reading or writing from a serial port).

Moving on from protected mode, we have the three different memory models: flat, segmented and real-address. Real-address mode is used at boot time and is kept around for backwards compatibility; however, it is quite likely you will never (or very rarely) encounter it and thus it's not discussed in this chapter. The flat memory model is about what you'd expect it to be from its name: it's flat (see Figure 2.2)! This means basically that the systems memory appears as a single contiguous address space, ranging from address 0 to 4294967296 in most instances. This address space is referred to as the linear address space, with any individual address being known as a linear address. In this memory model, all of the segments—code, stack, data, and so forth—fall within the same address space. Despite what you may be inclined to first think, this is the memory model used by nearly all modern OSes and chances are your computer's OS at home employs it. This seems like it would lead to disaster and unreliability; however, it is almost universally used with paging, which we'll discuss shortly.

Figure 2.2 Flat Memory Model

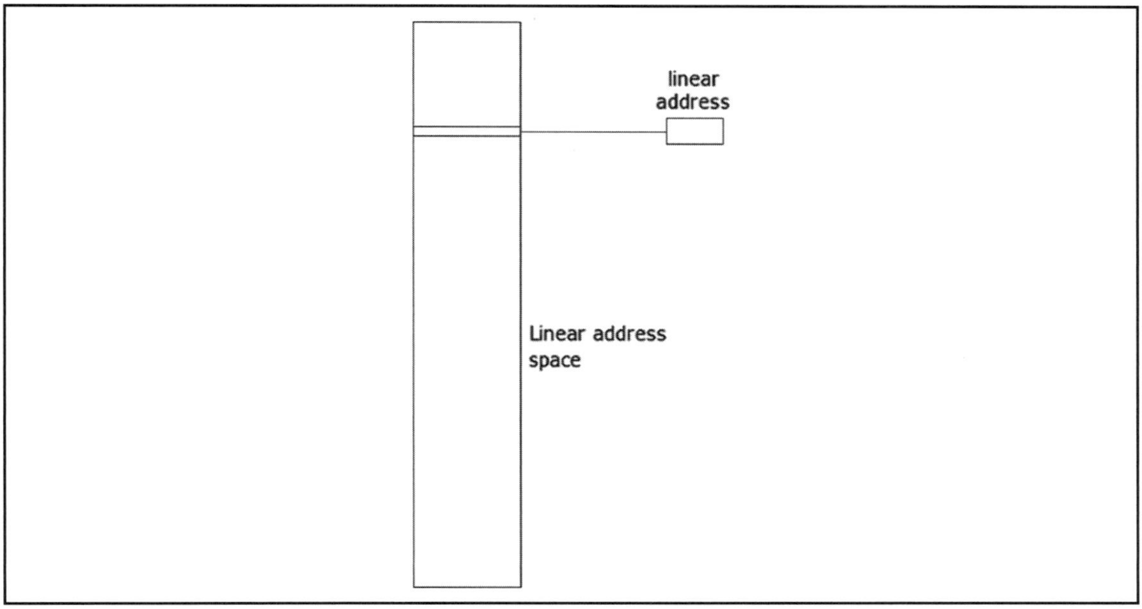

The flat memory model in protected mode differs only in that the segment limits are set to ensure that only the range of addresses that actually exist can be accessed. This differs from other modes, where the entire address space is used and there may be gaps of unusable memory in the address space.

The next memory model is the segmented memory model. It was used in earlier operating systems and has seen somewhat of a comeback, as in some arenas it implies an increase in speed (due to the ability to skip relocation operations) and security. Nonetheless, this memory model is still pretty rare and whether it makes a full comeback is yet to be seen, although it's a bit unlikely. It's described here because, if you do much reverse engineering or exploit development under Linux, your likelihood of encountering it goes up considerably.

In a segmented memory model, the systems memory is split up into subsections called segments. (See Figure 2.3.) These segments are kept independent and isolated from each other. To compare the segmented and flat memory models for a moment, what is really different between them is their representation to an OS or application. In both instances, the data is still stored in a linear fashion; however, the view of that data changes. With segmented memory, instead of issuing a linear address to access memory, a logical address (also known as a far pointer) is used. A logical address is the combination of a base, which is stored in a segment selector, and an offset. These two correspond to an address in that segment, which in turn maps to the linear address space. What this accomplishes is a higher degree of segmentation that is enforced by the processor, ensuring that one process does not run into another process (whereas in a flat memory model the same result is hopefully accomplished by the OS). Thus, the base address plus the offset equals a linear address in the processor's address space. Furthermore, a multiple segment model can be used that retains the same traits as a single segmented model, except each application is given its own set of segments and segment descriptors. At the present time, the author cannot think of an OS that employs this, so a further description of how it works is moot.

Figure 2.3 Segmented Memory Model

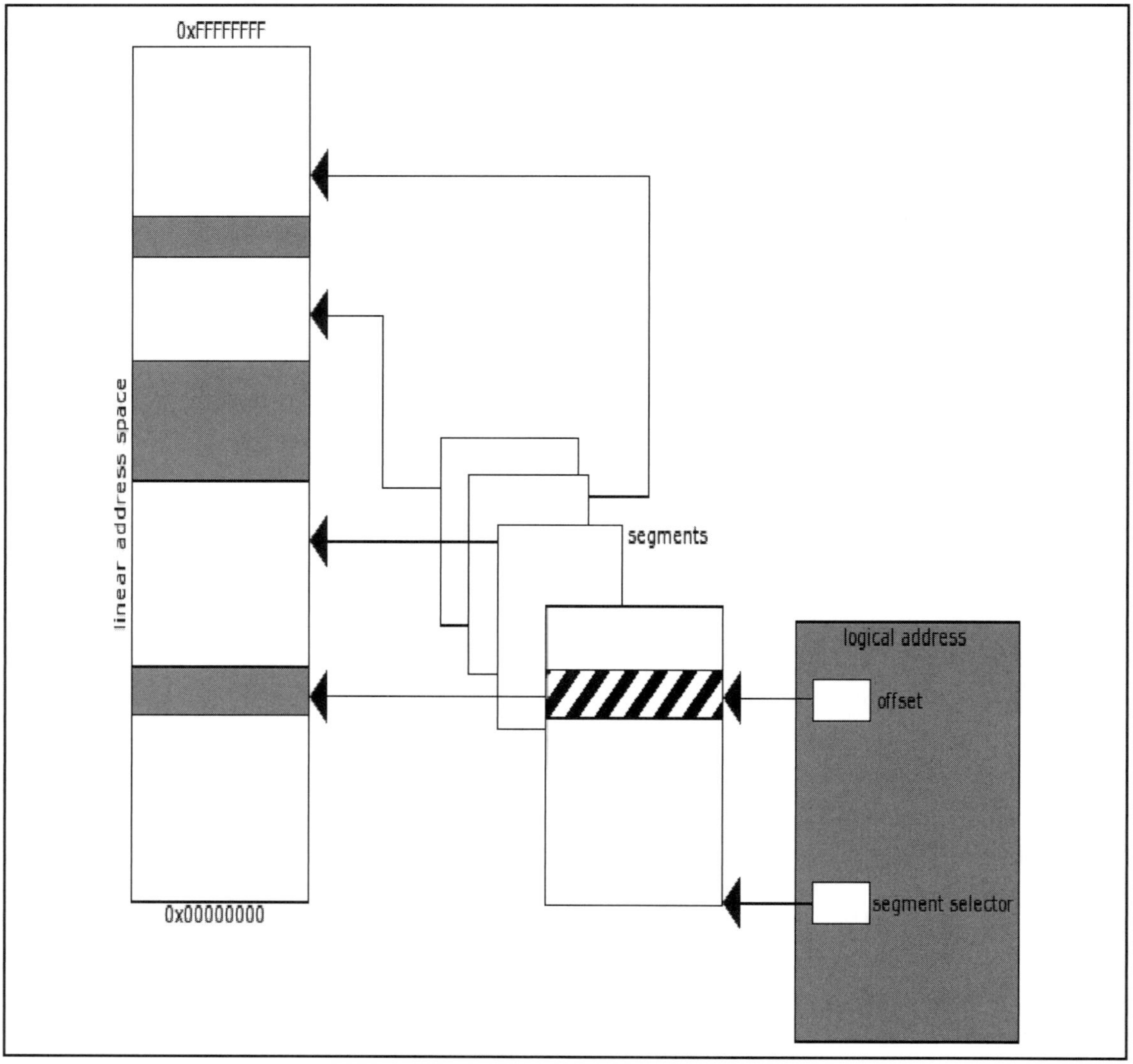

Damage and Defense...

The Security of Segmentation

As mentioned earlier, the segmented memory model has recently regained some traction in various communities, particularly in the case of grsecurity (http://grsecurity.com/) and PaX (http://pax.grsecurity.net/), which are third-party Linux kernel patches that provide superior security to that of the vanilla kernel. The lead develop of grsecurity, Brad Spengler, demonstrated the insecurity that a flat memory model can bring by product what is believed to be the first exploitable Linux kernel NULL pointer dereference, the details of which can be found at the following URL: http://marc.info/?l=dailydave&m=117294179528847&w=2. Furthermore, the anonymous author of PaX has implemented a feature called UDEREF which attempts to stop accidental dereferences of pointers provided by user-space pointers in the kernel (and thus a potentially exploitable condition). This feature has been documented in regards to how it works and the interested reader is encouraged to read the brief write-up to further understand the security implications of the flat memory model which UDEREF fixes. At the time of this writing, it can be found at the following URL: http://grsecurity.net/~spender/uderef.txt

As a result of modern OSes often using larger address space than physical memory can accommodate, some form of addressing all of the necessary data without requiring that it be stored in physical memory is required. The answer comes in the form of paging and virtual memory, one of the fundamental tenets of modern computing that is often misunderstood by people with a more operations-oriented background. It's not uncommon to encounter people who understand that a given application can be given 4 GB of memory but who don't really comprehend how that could possibly work, given that they only have 1 or 2 GB of physical memory.

In short, paging takes advantage of the fact that only the currently necessary data needs to be stored in physical memory at any given moment; it stores what data it needs in physical memory and stores the rest on disk drive. The process of loading memory with data from disk, or writing data to disk, is known as swapping data, and this is why Windows boxes typically have swap files and Linux boxes usually have swap partitions. When paging is not enabled, a linear address (whether it be formed from a far pointer or not) has a one-to-one mapping with a physical address and no translation between it and the issued address occurs. However, when paging is enabled, all pointers used by an application are virtual addresses. (This is why two applications in a protected mode flat memory model using paging accessing the exact same address don't trample each other.) These virtual addresses do not have a one-to-one mapping with a physical memory address.

When paging is employed, the processor splits the physical memory into 4 KB, 2 MB or 4 MB pages. When an address is turned into a linear address and then through the paging mechanism it is looked up, if the address does not currently exist in physical memory then a page-fault exception occurs, instructing the OS to load the given page into memory and then performing the instruction that generated the fault again.

The process of translating a virtual address to a physical address varies depending on the page size being used, but the basic concept is the same either way. A linear address is divided into either two or three sections. First the Page Directory Base Register (PDBR) or Control Register 3 (CR3) is used to locate a Page Directory. Following this, bits 22 through 31 in the linear address are used as an offset into the Page Directory which identifies a given Page Table to use. (See Figure 2.4.) Once the Page Table has been located, the bits 12 through 21 are used to locate a Page Table Entry (PTE), which identifies the page of memory to be used. Finally bits 0 through 11 of the linear address are used as an offset into the page to locate the data requested. When using other side pages, the process is nearly identical except that one layer of indirection is omitted; the directory entry points directly to the page and the page table and PTEs are completely omitted. The contents of Page Directory Entries (PDEs) and PTEs are not important for our purposes. If they become important to you at some point or if you're just curious, please refer to the processor's documentation.

Figure 2.4 4 KB Address Translation

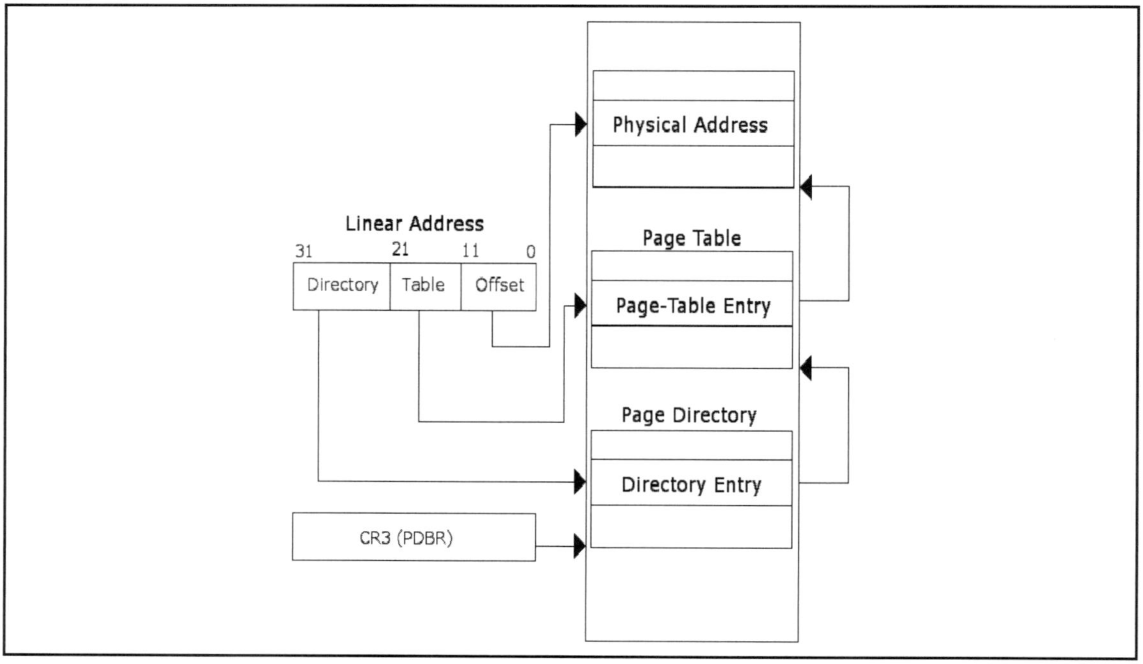

Now that we have a decent understanding of instructions and operands, memory models, operating modes and so on, we can move on. Most of the terms employed later on in this chapter and throughout the book have now been defined, so you can refer back to this section should you feel like you don't understand something as you work through the book.

The Stack, the Heap and Other Sections of a Binary Executable

In the previous section we talked some about segments, segment registers, segment descriptors and segment selectors, but we really didn't delve into the data that they contain. Understanding these various sections is fairly important to understanding the layout of a binary executable. In this section we will discuss these concepts as much as possible, although in some areas, such as with the heap, it's not really possible to jump into the depths of how it works exactly without a fairly in-depth analysis of a particular implementation; in those instances a generic high-level overview is provided and the understanding of the most minute details is left as an exercise for the reader.

WARNING

The reader should understand that sections and segments as defined here do not implicitly or explicitly imply a segmented memory model. In all of the memory models, applications and OSes are split into various sections; an application is blissfully unaware of the implementation details. Furthermore, throughout this section, the terms *segment* and *section* are used interchangeably.

Earlier we discussed the segment registers, in particular the CS, DS and SS segment registers, but we didn't tell you what the code, data and stack segments were. In traditional design, an application has a few different basic sections (and a lot of implementation-specific ones). The basic ones employed are the code segment (or text segment or simply .text), the data segment (often simply .data), the block started by symbol (BSS/.bss) segment, the stack segment and the heap segment. As an example, the following C code will help demonstrate the differences between the various sections:

```
unsigned int variable_zero;
unsigned int variable_one = 0x1234;
int
main(void)
{
        void * variable_two  = malloc(0x1234);
        [...]
}
```

In this code example, we have three variables defined and one function. The first variable, appropriately named *variable_zero*, is a variable of global scope that is uninitialized. In this instance, the C compiler will allocate space in the binary and fill it with zeros. The section of the binary it will exist in is the BSS section. The variable named *variable_one* is another globally scoped variable. However, in this instance it is initialized to the value 0×1234. In this instance, the compiler will preallocate the space for the variable in the binary and store the value in the data segment. Following this, we have the function *main*. *Main* obviously is a function, and thus is in the code segment. After *main* we find the variable called *variable_two*, which gives us an interesting predicament: we have a pointer whose scope is inside of *main* and the memory that it points to. The pointer itself is local to the function, is dynamically allocated and exists on the stack segment, giving it a lifetime of the function it exists in, whereas the pointer returned by *malloc()* exists on the heap, is dynamically allocated and has a global scope and a "use-until-free()" life expectancy. There are also often other sections; for instance, in programs compiled by the GNU compiler collection (GCC) a constant string that is declared in the source file will often end up in a section named *.rodata*, for read-only data, or in the code segment.

The stack is one of the more important sections to understand, as it plays a vital role in the routine operations of an application. Those with a formal computer science background will no doubt know what a stack is and how it works. A stack is a simple data structure that basically stacks data on top of each other and has elements added and removed in a last-in first-out (LIFO) manner. When you add an item to the stack, you push it onto a stack, and when you remove an item from a stack, you pop it off the stack. (See Figure 2.5.) The stack is important for two basic reasons on most computing platforms. The first is that all automatic or local variables are stored on it; that is, when a function is called, any local variables it has declared that are not static or similar have their space allocated on the stack. This is usually realized by adding or subtracting a number from the current stack pointer. The stack pointer is the ESP register, and it normally points to the top of the stack segment, or rather SS:ESP points to the top of the stack segment. The top of the stack is the lowest currently in use address of the stack—the lowest because the stack grows down on IA-32. The bottom of the stack is usually bounded, not by the absolute bottom, but the bottom of the current stack frame and is pointed to by the EBP register.

A stack frame is the current view into the stack relevant to the currently executing function; when the processor enters a new procedure, a few steps occur known as the procedure prologue. (See Figure 2.6.) The procedure prologue is as follows: the routine first pushes the address of the

Figure 2.5 A Stack

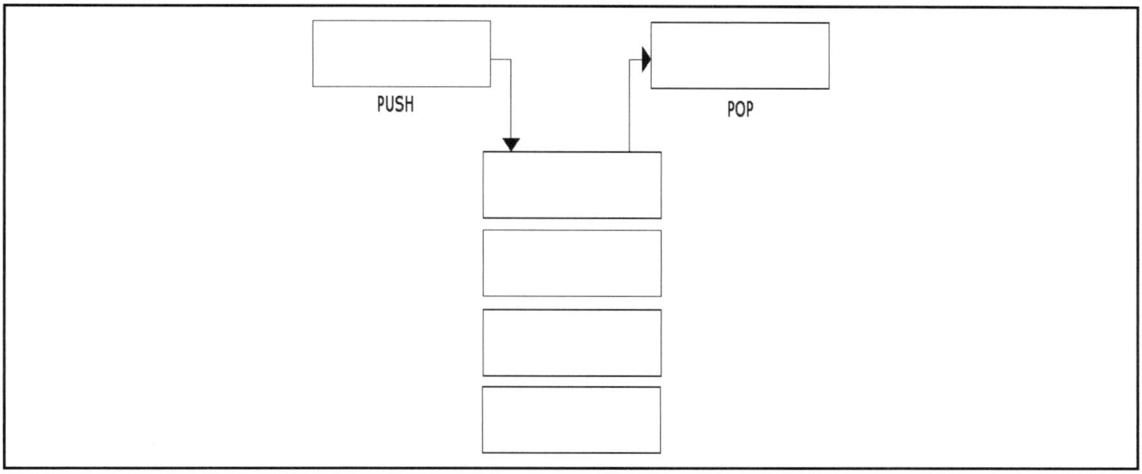

next instruction in the calling frame onto the stack; next the current base (EBP) of the stack is saved onto the stack; the ESP register is then copied into the EBP register; and finally the ESP register is decremented to allocate space for variables in that function. When the function is called, the arguments to the function are pushed onto the stack in reverse order (or the last one is pushed first). The assembly generated for the prologue is as follows:

```
push ebp
mov ebp, esp
sub esp, 0x1234
```

Figure 2.6 Stack Frame

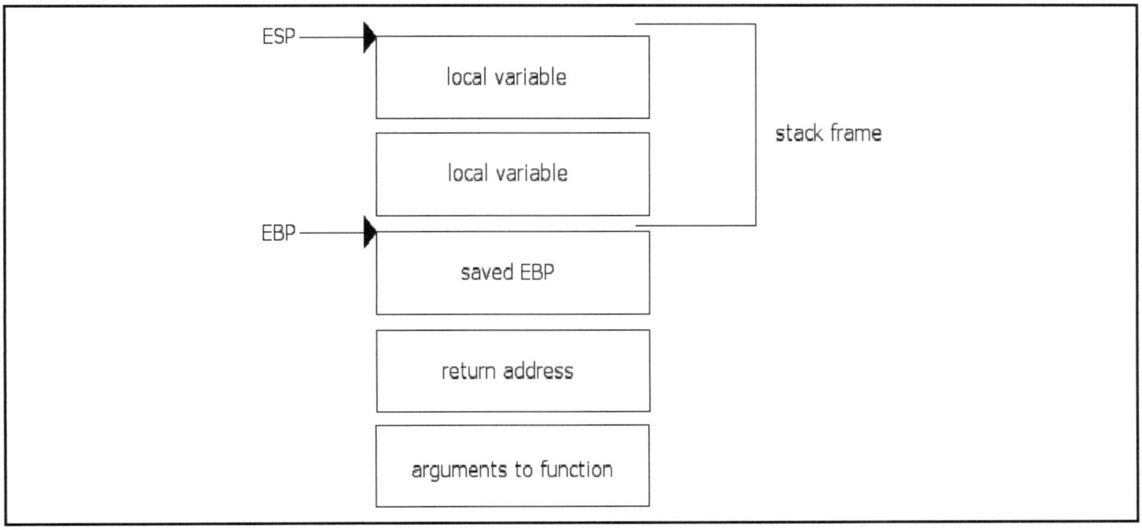

What we see here is strangely missing the saving of the return address, or the address we will continue execution at in the calling routine. For instance, given the following C code:

```
A();
B();
```

when inside of the function *A()* the return address would be the address of the instruction *B()*. It's a little tricky to understand at first, especially because you don't ever see the instruction that saves the address onto the stack, but rather it is implied by using the call instruction, which is discussed a little later on in the chapter. In addition to the procedure prologue, there is also the procedure epilogue. The procedure epilogue basically undoes everything that the prologue did. This includes incrementing the stack pointer to deallocate any local variables, and then calling the *leave* and *ret* instructions, which remove the saved frame pointer and return address and return execution flow to the calling function. So, to summarize the points on the stack:

- The stack grows down, towards lower addresses on IA-32.
- Items are removed and added onto a stack in LIFO order.
- Variables that are locally scoped and only exist for the lifetime of the function end up on the stack.
- Each function has a stack frame (unless specifically omitted by the compiler) that contains the local variables.
- Prior to each function's stack frame there is a saved frame pointer, return address and the parameters to the routine.
- The stack frame is constructed during the procedure prologue and destructed during the procedure epilogue.

The *heap* is another important data structure, but not because any features of the processor depend on it, but rather because of the large amount of use it receives. The heap is simply a section of memory that is used for dynamically allocated variables that need to exist outside of the current stack frame; as a result of trait, most of the objects and indeed large amounts of the data an application uses will be on the heap. The heap is usually either the result of a random mapping or in more classic examples it was a dynamic extension of the data segment (although DS rarely if ever points to the heap). In that sense, the processor is by and large ignorant and the details are hidden away from the processor. Furthermore, the OS knows very little about the user-land heap; when requested, it simply gives the application more memory if possible and fails otherwise. It is typically the libc or similar that provides the heap operations and thus defines its semantics.

The heap, typically upon initialization, will request a fairly large section of memory from the OS, and will hand out smaller chunks of memory based upon requests from the application. These chunks will typically have inline metadata indicating the chunk's size and other elements, such as the size of the previous block of memory.

The blocks of allocated memory are navigated by taking the pointer to a given chunk and adding its size to it to find the next chunk, or by subtracting the previous size from the beginning of the chunk to find the previous one. For instance, in Figure 2.7 you will find an example of an allocated chunk as represented in Glibc. In this instance the pointer labeled *mem* indicates that start of data returned to the API user by *malloc()* or similar, whereas the pointer labeled *chunk* marks the beginning of the actual chunk. There we find that there is metadata including the size of the previous chunk and the size of the current chunk, along with flags indicating various status conditions. This chunk, while mostly used by Linux, is generically similar to chunks used by most operating systems and dynamic memory allocator implementation (with, of course, some key differences). Since initially obtaining that large chunk of memory from the OS or extending the size of the data segment are fairly expensive operations, a cache of sorts is usually maintained. This cache usually comes in the form of a linked list of pointers to previously free()'d chunks of memory. This list is typically fairly complex, with the blocks of memory being coalesced into adjacent free blocks of memory to reduce fragmentation, and with various lists sorted by size or some other characteristic to allow the most efficient means possible of locating a candidate block of memory whenever an allocation request occurs.

Figure 2.7 Glibc Allocated Chunk

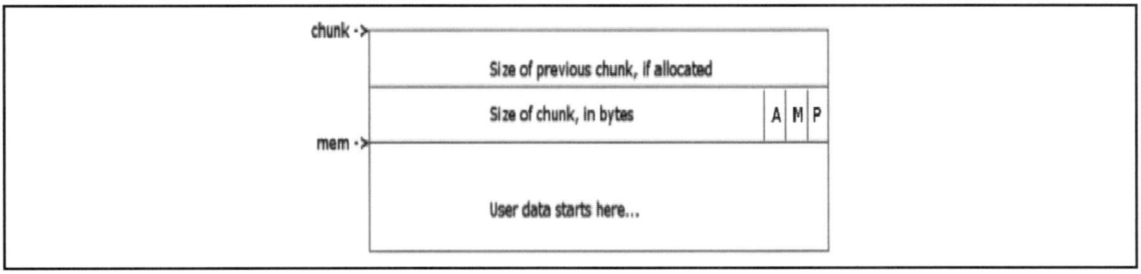

To use a similar example to the previous one, in Figure 2.8 you will see the representation of a free block of memory as represented by Glibc. In this instance, the pointer labeled *mem* indicates where the pointer returned to the API user used to be, and the one labeled *chunk* points to the beginning of the physical data structure used. The biggest difference is that in what used to be user data, there are now two pointers stored pointing to the next free block of memory in the linked list and the previous block of memory in the linked list. This of course implies that, unlike allocated chunks which are navigated by size, a free block of memory is navigated directly by linked list. The specific details of the structure listed are, again, specific to Glibc; however, the concept itself is generic enough to apply to most implementations. Thus, as allocation and free requests come in and out, which happens quite frequently throughout the lifetime of your average application, chunks are taken away from the original chunk of memory obtained from the OS and returned to free lists, and then if possible further allocation requests make use of these blocks of memory on the free list, and so on until either all memory is in use, or the application terminates.

Figure 2.8 Glibc Free Chunk

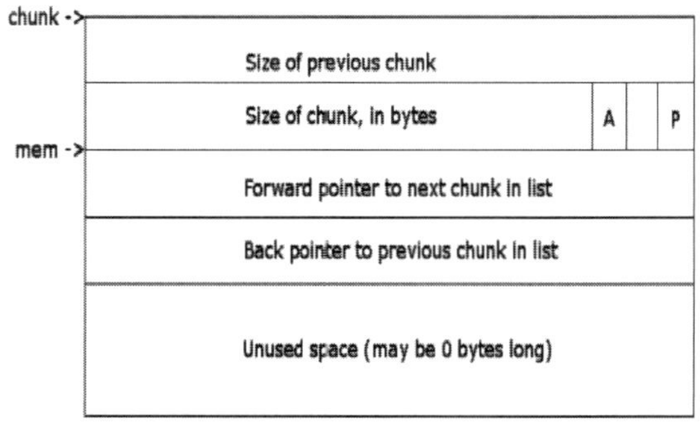

In the previous section, we've discussed the most common sections of a binary executables layout, including some of their functions, and took a more in-depth tour of the stack segment and the heap segment and talked about how they worked to some degree. This should be enough to provide a base to continue building your understanding. Of course, the interested reader is encouraged to refer to other works more specifically targeted at questions they may have on these subjects.

IA-32 Instruction Set Refresher and Reference

In the prior sections, we talked briefly about instructions and operands, but focused more on the architectural design of IA-32 and then delved into some common layouts for binary executable memory and their purposes and uses. In this section the intention is to provide you with a reference for some of the more commonly used instructions and talk to you some about their uses and operands. If by now you've already read the Intel developer manuals, then this section is likely to be redundant and you may wish to skip directly to the next chapter. In this section the terminology shown in Table 2.1 will be used.

Table 2.1 Terminology Employed

Term	Meaning
Reg32	Any 32-bit register
Reg16	Any 16-bit register
Reg8	Any 8-bit register
Mem32	32-bit memory operand
Mem16	16-bit memory operand
Mem8	8-bit memory operand
Sreg	Segment register
Memoffs8	8-bit memory offset
Memoffs16	16-bit memory offset
Memoffs32	32-bit memory offset
Imm8	8-bit immediate (constant)
Imm16	16-bit immediate
Imm32	32-bit immediate
ptr16:16	absolute address given in operand
ptr16:32	absolute address given in operand
mem16:16	absolute indirect address given in mem16:16
mem16:32	absolute indirect address given in mem16:32
rel8	8-bit relative displacement
rel16	16-bit relative displacement
rel32	32-bit relative displacement
Register name	Any name of a register that has already been introduced

The first instruction we are going to cover is the *mov* instruction, which is a very basic instruction that copies one operand to the other. It can take the forms and allows the operands shown in Table 2.2.

Table 2.2 *mov* Instruction

Destination Operand	Source Operand
reg8/mem8	reg8
reg16/mem16	reg16
reg32/mem32	reg32
reg8	reg8/mem8
reg16	reg16/mem16
reg32	reg32/mem32
reg16/mem16	Sreg
Sreg	reg16/mem16
AL	memoffs8
AX	memoffs16
EAX	memoffs32
memoffs8	AL
memoffs16	AX
memoffs32	EAX
reg8	imm8
reg16	imm16
reg32	imm32
reg8/mem8	imm8
reg16/mem16	imm16
reg32/mem32	imm32

The *mov* instruction copies the source operand to the destination operands and can only be used to move certain types of operands; for instance, it cannot be used to set the Code segment, and it cannot be used to modify the EIP register. If a destination operand is of type *Sreg,* then it must point to a valid segment selector. The next instructions introduced will be the various bitwise operations such as *and* and *exclusive or.*

The *and* instruction is another fairly simple instruction. It performs a bitwise AND on the destination operand with the source operand, storing the result in the destination operand. It supports the operands shown in Table 2.3. A bitwise AND compares the binary representation of the two operands, and sets the output bit to 1 if both bits compared are turned on; otherwise the result is a turned-off bit.

Table 2.3 *and* Instruction

Destination Operand	Source Operand
reg8/mem8	reg8
reg16/mem16	reg16
reg32/mem32	reg32
reg8	reg8/mem8
reg16	reg16/mem16
reg32	reg32/mem32
AL	imm8
AX	imm16
EAX	imm32
reg8	imm8
reg16	imm16
reg32	imm32
reg8/mem8	imm8
reg16/mem16	imm16
reg32/mem32	imm32

The next instruction referenced is the *not* instruction, another fairly simple but commonly used instruction; it performs a bitwise NOT operation on a single operand and allows the operands shown in Table 2.4. This simply sets each 1 to 0 and vice versa in its operand.

Table 2.4 *not* Instruction

Destination Operand	Source Operand
reg8/mem8	N/A
reg16/mem16	N/A
reg32/mem32	N/A

Chugging forward, we have the *or* instruction, which performs a bitwise OR on its arguments and takes the arguments shown in Table 2.5. A bitwise OR is, again, a fairly simple operation that's used often. It compares the two operands bit by bit and sets the corresponding output bit to zero only if both compared bits are set to zero.

Table 2.5 *or* Instruction

Destination Operand	Source Operand
reg8/mem8	reg8
reg16/mem16	reg16
reg32/mem32	reg32
reg8	reg8/mem8
reg16	reg16/mem16
reg32	reg32/mem32
AL	imm8
AX	imm16
EAX	imm32
reg8	imm8
reg16	imm16
reg32	imm32
reg8/mem8	imm8
reg16/mem16	imm16
reg32/mem32	imm32

Next in the line-up we have the exclusive-or instruction, or *xor*. It performs a bitwise exclusive-or (XOR) on its operands and takes the operands shown in Table 2.6. The *xor* instruction compares the source and the destination operands and stores the output in the destination operand. Each output bit is 1 if the two compared bits are different; otherwise the output bit is 0.

Table 2.6 *xor* Instruction

Destination Operand	Source Operand
reg8/mem8	reg8
reg16/mem16	reg16
reg32/mem32	reg32
reg8	reg8/mem8
reg16	reg16/mem16
reg32	reg32/mem32
AL	imm8

Continued

Table 2.6 Continued

Destination Operand	Source Operand
AX	imm16
EAX	imm32
reg8	imm8
reg16	imm16
reg32	imm32
reg8/mem8	imm8
reg16/mem16	imm16
reg32/mem32	imm32

The *test* instruction is commonly used to determine a specific condition and then modify control flow based on the results (see Table 2.7). The *test* instruction performs a bitwise AND of the first and second operands, and then sets flags in the EFLAGS register accordingly. Following this, the result is then discarded.

The *cmp* instruction compares two operands; this comparison is performed by subtracting the source operands from the destination operands and setting flags in the EFLAGs register

Table 2.7 *test* Instruction

Destination Operand	Source Operand
reg8/mem8	reg8
reg16/mem16	reg16
reg32/mem32	reg32
AL	imm8
AX	imm16
EAX	imm32
reg8	imm8
reg16	imm16
reg32	imm32
reg8/mem8	imm8
reg16/mem16	imm16
reg32/mem32	imm32

accordingly (see Table 2.8). It is often used in a manner similar to the test instruction and is used to compare values like user input and return values from routines. If an immediate value is used as an operand, it is sign-extended to match the size of the other operand.

Table 2.8 *cmp* Instruction

Destination Operand	Source Operand
reg8/mem8	reg8
reg16/mem16	reg16
reg32/mem32	reg32
reg8	reg8/mem8
reg16	reg16/mem16
reg32	reg32/mem32
AL	imm8
AX	imm16
EAX	imm32
reg8	imm8
reg16	imm16
reg32	imm32
reg8/mem8	imm8
reg16/mem16	imm16
reg32/mem32	imm32

The *load effective address* instruction, or *lea*, calculates the address as specified by the source operand and stores it in the destination operand (see Table 2.9). It is also used as a method for doing arithmetic between multiple registers without modifying the source operands.

Table 2.9 *lea* Instruction

Destination Operand	Source Operand
reg8	mem8
reg16	mem16
reg32	mem32

The *jmp* instruction transfers execution control to its operand. This instruction can execute four different types of jumps: a near jump, a short jump, a far jump and a task switch (see Table 2.10). A near jump is a jump that occurs within the current code segment. A short jump is a jump to an address within −128 to 127 from the current address. A far jump can take control to any segment in the address space providing it is of the same privilege level as the current code segment. Finally, a task switch jump is a jump to an instruction in a different task.

Table 2.10 *jmp* Instruction

Destination Operand	Source Operand
rel8	N/A
rel16	N/A
rel32	N/A
reg16/mem16	N/A
reg32/mem32	N/A
ptr16:16	N/A
ptr16:32	N/A
mem16:16	N/A
mem16:32	N/A

The *jcc* instructions are not one particular instruction, but rather a series of conditional jumps. The conditions vary with the instruction used, but they're typically used in collaboration with the *test* and *cmp* instructions. Table 2.11 shows destination operands for *jcc* instructions. In Table 2.12 you will find a list of conditional jumps and the flags that they check to determine whether a condition is true or not. Don't worry about the flags just yet; the description of the EFLAGS register will come just after the instructions.

Table 2.11 *jcc* Instructions

Destination Operand	Source Operand
rel8	N/A
rel16	N/A
rel32	N/A

Table 2.12 Conditional Jump Registers

Instruction	EFLAGS condition	Description
ja	*CF = 0 && ZF = 0*	Jump if above
jae	*CF = 0*	Jump if above or equal
jb	*CF = 1*	Jump if below
jbe	*CF = 1 || ZF = 1*	Jump if below or equal
jc	*CF = 1*	Jump if carry
jcxz	*CX = 0*	Jump if CX is zero
jecxz	*ECX = 0*	Jump is ECX is zero
je	*ZF = 1*	Jump if equal
jg	*ZF = 0 && SF = OF*	Jump if greater than
jge	*SF = OF*	Jump if greater than or equal to
jl	*SF != OF*	Jump if less than
jle	*ZF = 1 || SF != OF*	Jump if less than or equal to
jna	*CF = 1 || ZF = 1*	Jump if not above
jnae	*CF = 1*	Jump if not above or equal
jnb	*CF = 0*	Jump if not below
jnbe	*CF = 0 && ZF = 0*	Jump if not below or equal
jnc	*CF = 0*	Jump if not carry
jne	*ZF = 0*	Jump not equal
jng	*ZF = 1 || SF != OF*	Jump not greater
jnge	*SF != OF*	Jump not greater or equal
jnl	*SF = OF*	Jump not less
jnle	*ZF = 0 && SF = OF*	Jump not less or equal
jno	*OF = 0*	Jump if not overflow
jnp	*PF = 0*	Jump not parity
jns	*SF = 0*	Jump not signed
jnz	*ZF = 0*	Jump not zero
jo	*OF = 1*	Jump if overflow
p	*PF = 1*	Jump if parity
jpe	*PF = 1*	Jump if parity even
jpo	*PF = 0*	Jump if parity odd
js	*SF = 1*	Jump if signed
jz	*ZF = 1*	Jump if zero

So, as you can see in Table 2.12, there are quite a few conditional jump registers, and all of them depend on the various states of the EFLAGS register, which we haven't really described yet, but will do now. The EFLAGS register is a 32-bit register that contains a group of status and system flags, and a control flag. Each flag is represented by a single bit in this register, and moving from bit 0 to 31 we have the following flags:

CF Carry flag, indicates a carry or a borrow out of the most significant bit of a register in an arithmetic operation. This flag is used in unsigned arithmetic.

PF Parity flag, set if the least significant bit of the result contains an even number of bits turned on.

AF Adjust flag, set if an operation resulted in a carry or borrow from bit 3.

ZF Zero flag, set if the result of an operation is zero.

SF Sign flag, set to the most significant bit of a result (which is the sign bit in a signed integer)

TF Trap flag, set to enable single-stepping debugging or when single-stepping debugging is enabled.

IF Interrupt enable flag, set when maskable interrupts are enabled, cleared when they are blocked.

DF Direction flag, used in string operations to determine whether the instructions increment or decrement

OF Set if the result of an operation resulted in an integer overflow when performing signed arithmetic.

IOPL flags (bits 12 and 13) I/O privilege level flags, indicates the current I/O privilege level of the currently running task

NF Nested task flag, set when the current task is associated with the previously executed task

RF Resume flag, controls the processor's response to debug exceptions

VM Virtual-8086 flag, set to enable virtual 8086 mode

AC Alignment check flag, set to enable alignment checking of memory references

VIF Virtual interrupt flag, virtual image of the IF flag, used together with the VIP flag

VIP Virtual interrupt flag, used to determine if an interrupt is pending

ID Identification flag, used to determine if the CPU supports the CPUID instruction.

Bits 22 through 31 are currently reserved.

The *call* instruction is somewhat akin to a more official jump; it sets up the stack as previously described in a manner that will allow the processor to resume execution in the calling function when the executed function finishes. (See Table 2.13.)

Table 2.13 *call* Instruction

Destination Operand	Source Operand
rel16	N/A
rel32	N/A
reg16/mem16	N/A
reg32/mem32	N/A
ptr16:16	N/A
ptr16:32	N/A
mem16:16	N/A
mem16:32	N/A

The *ret* instruction is the inverse of the *call* instruction. It takes the metadata stored on the stack, pops it off and returns to that address (see Table 2.14). The optional immediate operand specifies how many bytes to pop off of the stack after performing the return.

Table 2.14 *ret* Instruction

Destination Operand	Source Operand
N/A	N/A
imm16	N/A

As you can see, there are already a lot of instructions to be familiar with, and most of them take many different operands, which in turn results in many different forms of the same instruction (and thus different opcodes). This is hardly a complete instruction reference—in fact, it barely scratches the surface. However, we wanted to touch base on some of the more commonly used instructions and at least familiarize you with them. Readers are strongly encouraged to consult the Intel developer manuals, specifically 3A and 3B, if they are not already familiar with the instruction set.

Summary

So now you're at least familiar with assembly and the Intel architecture. You know how the memory models and operating modes work to some degree and have a decent general base of comprehension to build on. However, you might find yourself thinking, "Yes, I understand some assembly now, but what is reverse engineering?". Well, reverse engineering is a broad term and means different things to different people. In general, the answer is that it's the process of taking an application that is in a form not meant for human readability or analysis and working backwards towards the beginning. Some people do this in order to regain lost source code, others do it to duplicate proprietary products, some reverse malicious applications to know what they do, and finally others reverse engineer software to find bugs and vulnerabilities in it.

To be brief, and for the purposes of this book, reverse engineering is taking a binary file that is meant to be read by the computer and using the opcodes to generate assembly, and then reading that assembly to help accomplish whatever goals you may have. In this sense, it can be said that, to the reverse engineer, there is no such thing as closed source software. The processor sees every instruction, and so does the reverse engineer.

Chapter 3

Portable Executable and Executable and Linking Formats

Solutions in this chapter:

- **Portable Executable Format**
- **Executable and Linking Format**

☑ **Summary**

Introduction

In this chapter we will introduce two common binary executable formats, Portable Executable (PE) and Executable and Linkable Format (ELF). PE is the binary format used in Windows, while ELF is used by many of the Unices. ELF is a replacement for the older a.out format that did not include standardized support for shared libraries. Furthermore, PE is an offshoot of the COFF format, which was used in an earlier Unix, and the author's understanding is that this is by and large the result of many of the developers being hired at Microsoft.

At any rate, this chapter serves as an introduction to the physical layout of the files, and details aspects of the files that a reverse engineer would find interesting and/or important. Both of these file formats have open documentation and, in places where readers find this chapter lacking, they are strongly encouraged to read the specifications themselves.

Portable Executable Format

Portable Executable format, more correctly termed the Portable Executable and Common Object File Format (PE-COFF), is a fairly simple format that is easily understood. Here we will cover the absolute basics and try to avoid covering topics that were defined in the previous section (for instance, little attention is given to the .text or code section, or .data section). Instead we'll focus on getting the reader up to speed on being able to open a file in a hex editor and navigate to the various headers of the file. With the exception of a few important sections, the internal format is not defined, with this left as an exercise for the reader.

In Figure 3.1 you will find a basic diagram dictating the general layout of a typical PE file. At the beginning you will find a DOS stub program, with a header. This is a simple program designed to be run if the application is run in DOS; it is used for backwards compatibility. To the reverse engineer, the only important part of this is that it contains an offset to the PE header. This can be found by simply seeking to the offset 0x3C from the beginning of the file. This offset in turn dictates an offset relative to the beginning of the file where the PE header can be found.

It should be noted that the DOS header itself is only found in image files and not in other files such as objects and so on. One characteristic of the DOS file header is that the first few bytes will contain the "magic" bytes indicating that it is a DOS image, specifically the characters M and Z. This knowledge, combined with other simple heuristics, can help a reverser identify a file as being a PE. Skipping past the DOS header information, as it is essentially useless to us, we find the PE header itself. The PE header begins with another

Figure 3.1 Typical PE Layout

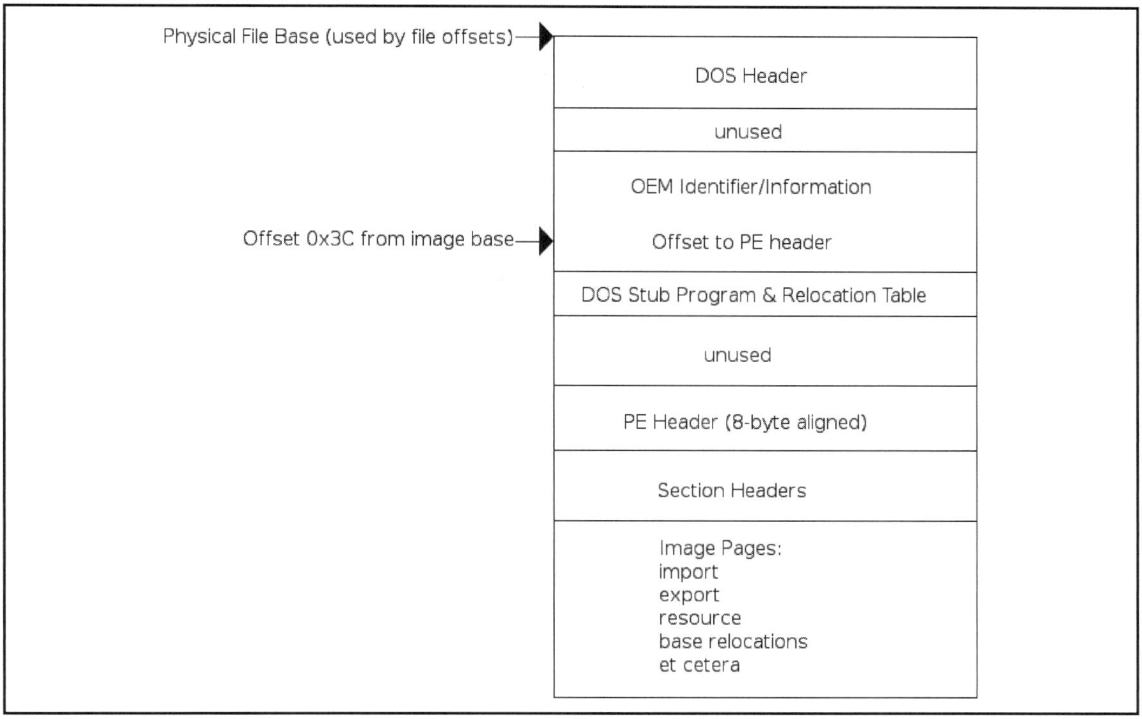

field that identifies it as a PE file, specifically the bytes PE\0\0, where \0 is a binary zero. Immediately following this signature, there is the COFF file header (see Figure 3.2). The first field in the format is the machine type; it is two bytes long and will almost always be either 0x8664 (AMD64), 0x14c (IA32) or less often 0x200 for IA64 Itanium processors. Immediately following this field is one that indicates the number of sections in the file. It is also a two-byte field, and, according to Microsoft documentation, the maximum value this can hold is 96. Skipping along to the next item of interest, we have a file offset to the COFF symbol table. As a reverse engineer, you can almost guarantee that the file you're inspecting will not have symbols, so you can expect that this field will be zero, indicating that an offset does not exist. However, if an offset *does* exist, it will be specified here. After this field, there is another four-byte field indicating the number of symbols present. Finally, the last two standard COFF fields are both two bytes, one indicating the size of the optional header (which exists in images only) and a field called *characteristics* that defines specific attributes of the file.

Figure 3.2 COFF File Header

Offset	Size	Field Name
0	2	**Machine**
2	2	**Number Of Sections**
4	4	Time Date Stamp
8	4	**Pointer To Symbol Table**
12	4	**Number Of Symbols**
16	2	**Size Of Optional Header**
18	4	**Characteristics**

Bold incicates discussed/referenced field

The characteristics specify various attributes of the file, such as whether the file is a dynamically loaded library (DLL) or not, if the file is part of the system, if the file has had its relocation information stripped, if the file uses 32-bit words, and so on. For a full table describing this information, please consult the official Microsoft documentation.

The optional header, if one exists (it does not exist in object files), is broken into three major sections, the first being eight fields that are generic to the COFF format, followed by 21 Windows-specific fields, and finally the data directories. (See Figure 3.3.) In the generic COFF section, the fields that we would find most interesting are the *magic*, *size of code*, *size of initialized data*, *size of uninitialized data*, *entry point and code base address*, and *data base address* fields. The *magic* field is another signature that can be used to identify PE versions; the valid fields are 0x10B and 0x20B for PE32 and PE32+, respectively. This chapter will cover only the PE32 format, as it is more common, and will leave the PE32+ format as yet another exercise for the reader. The size fields are relatively self-explanatory and indicate the size of the .text/code, .data and .bss sections of the file. Finally, we have the *relative virtual address* (RVA) of the entry point, or place where the application should start executing. An RVA is just a complicated way of referring to an offset from the virtual address (VA) where the file is loaded, and should be looked at as such. Finally, after all of the above, we have the base virtual addresses of the code and data sections. Next, in the

Windows-specific section of the optional header, we find the following fields of interest: image base, size of image, size of headers, DLL characteristics, and finally the number of data-directory entries and their size.

Figure 3.3 Optional Header Layout

Offset	Size	Optional Header Section
0	2	Standard COFF fields
28	68	Windows Specific fields
96	Variable	Data directories

Bold indicates discussed/referenced field

The image base specifies the preferred address of the first byte of the image when loaded. According to Microsoft documentation, it must be a multiple of 64 K. The default varies by operating system and is not really important; it's merely a preference thing and the loader has the option of overriding what the file thinks is right. Next, there is the size of the image and the size of the headers; the size of the image is the total size of the file including all of the headers as loaded into memory, with the size of the headers specifying the size of the headers, obviously. The DLL characteristics field is obviously used only for DLLs and specifies attributes specific to the DLL. To the reverser, the interesting fields are 0x0040, which specifies that the base address can be assigned dynamically and would allow for things such as address space layout randomization (ASLR), 0x0080 which specifies that code integrity checks are made, 0x0100 which specifies that the image is no-execute (NX) compatible, and finally 0x0400 which indicates that the file does not use *structured exception handling* (SEH), effectively preventing any SEH handler from pointing into this DLL. Finally, the last element of interest specifies the number of elements in the next subsection of the optional header. See Figure 3.4.

Figure 3.4 Optional Header

Offset	Size	Field Name	Group
0	2	**Magic**	COFF Standard fields
2	1	MajorLinkerVersion	COFF Standard fields
3	1	MinorLinkerVersion	COFF Standard fields
4	4	**SizeOfCode**	COFF Standard fields
8	4	**SizeOfInitializedData**	COFF Standard fields
12	4	**SizeOfUnitializedData**	COFF Standard fields
16	4	**EntryPoint**	COFF Standard fields
20	4	**BaseOfCode**	COFF Standard fields
24	4	**BaseOfData**	COFF Standard fields
28	4	**ImageBase**	Windows Specific fields
32	4	SectionAlignment	Windows Specific fields
36	4	FileAlignment	Windows Specific fields
40	2	MajorOSVersion	Windows Specific fields
42	2	MinorOSVersion	Windows Specific fields
44	2	MajorImageVersion	Windows Specific fields
46	2	MinorImageVersion	Windows Specific fields
48	2	MajorSubsystemVersion	Windows Specific fields
50	2	MinorSubsystemVersion	Windows Specific fields
52	4	Win32VersionValue	Windows Specific fields
56	4	**SizeOfImage**	Windows Specific fields
60	4	**SizeOfHeaders**	Windows Specific fields
64	4	Checksum	Windows Specific fields
68	2	Subsystem	Windows Specific fields
70	2	**DLLCharacteristics**	Windows Specific fields
72	4	SizeOfStackReserve	Windows Specific fields
76	4	SizeOfStackCommit	Windows Specific fields
80	4	SizeOfHeapReserve	Windows Specific fields
84	4	SizeOfHeapCommit	Windows Specific fields
88	4	LoaderFlags	Windows Specific fields
92	4	**NumberOfRVAsAndSizes**	Windows Specific fields

Bold indicates discussed/referenced field

Data directories are a bit of a different beast. They specify some other type of data for the image. For instance, the import table data directory specifies which libraries and functions will be imported by the application for use. The data directory section of the optional header is an array of 16 structures containing double word values, specifying the virtual address and size for a given data directory. The data directories specified are shown in Table 3.1.

Table 3.1 Data Directories

Name	Description
Export table	Export table specifies functions exported by the file
Import table	Import table specifies functions imported by the file
Resource table	Resource table specifies various resources used by the file, such as icons
Exception table	Exception table specifies registered exception handlers used by the file
Certificate table	Attribute certificate table
Base relocation table	Base relocation table specifies all base relocations in the file
Debug	Used for storing compiler-generated debugging information
Architecture	Reserved, must be zero
Global pointer	RVA of the value to be stored in the global pointer register
Thread local storage (TLS) table	Specifies information used in thread-specific data storage
Load configuration table	Different uses for different Windows versions, since XP used to register SafeSEH functions
Bound import	Bound import table
Import address table (IAT)	Prior to runtime, identical import lookup table, at runtime filled with resolved symbol addresses
Delay import descriptor	Similar to the import table but delays imports
CLR runtime header	CLR runtime header
Reserved	Reserved, must be zero

At this point, only the export, import and load configuration tables are described in detail. Everything else is left as an exercise for the reader. The export table, as previously indicated, specifies the functions exported by the file. The format of the export table, also known as the .edata section, is shown in Table 3.2.

Table 3.2 Format of the Export Table

Name	Description
Export directory table	The export directory table describes the entirety of the export information. It contains address information that is used to resolve imports to the exported functions within the image.
Export address table	Contains the address of exported entry points, data and absolutes
Name pointer table	Array of RVAs into the export name table
Ordinal table	Array of 16-bit ordinals into the export address table
Export name table	Null-terminated variable length string names of exported functions/data/etc.

It should be noted that not all of these tables are required to be present; if exports are only to be done via ordinal, then only the export directory table and export address table are required. The interesting fields of the export directory table (EDT) are: the *name RVA*, *ordinal base*, *address table entries*, *number of name pointers*, *export address table* (EAT) *RVA*, *name pointer RVA*, and *ordinal table RVA*. See Figure 3.5.

The *name RVA* is the RVA to the name of the DLL in question. The *ordinal base* is simply the base index that ordinal indexing starts from; this is typically set to one. The *address table entries* and *number of name pointers* fields specify how many entries there are in the address table and name table, respectively. The *EAT RVA*, *name pointer RVA* and *ordinal table RVA* entries are all exactly what they sound like; they indicate the RVA for the rest of the tables in the .edata section. The *export address table* is a fairly simple structure with only one element that can be represented one of two ways, and is most likely implemented as a union. If the address is not within the export section (which is defined by summing the address and length as provided in the optional header), the field is an actual address in the code or data. Otherwise the field is a forwarder RVA, which names a symbol in another DLL. The *export name pointer table* is another simple structure; it simply contains an RVA into the export name table for each export, if defined. An export name is only defined if a pointer is contained in

Figure 3.5 EDT

Offset	Size	Field Name
0	4	Export Flags
4	4	Time/Date Stamp
8	2	Major Version
10	2	Minor Version
12	4	**Name RVA**
16	4	**Ordinal Base**
20	4	**Address Table Entries**
24	4	**Number Of Name Pointers**
28	4	**Export Address Table RVA**
32	4	**Name Pointer RVA**
36	4	**Ordinal Table RVA**

Bold indicates discussed/referenced field

this table. The *ordinal table* is an array of 16-bit indexes biased by the ordinal base into the EAT. The ordinal table and the export name table are essentially mirrors of each other in that an index into one provides an index into the other, providing of course that the name for the export exists. See Figure 3.6.

Figure 3.6 EAT

Offset	Size	Field Name
0	4	Export RVA
0	4	Forwarder RVA

Finally, the *export name table* contains the actual string that makes up the public name for the exported symbol; *public* means that if one exists, an application can import the function/data by name. So, taking all of this into account, a symbol can be resolved by name using the following steps:

0. Obtain the VA or the export directory table in the optional header.
1. Use that VA to locate the ordinal base, export directory table and the ordinal table RVAs.
2. Retrieve the RVA of the name pointer RVA.
3. Search the export name pointer table to determine if the function is exported by name.
4. Use the index into the name pointer table as an index into the ordinal table to retrieve the ordinal.
5. Take the ordinal and subtract it from the ordinal base and use the result as an index into the EAT.
6. The data at this index is the RVA for the exported function.

The process for obtaining a symbol via ordinal is exactly the same. However, the steps for finding the export by name are removed and the conversion from name pointer index to ordinal index is also removed.

The import table, or .idata section, uses a method similar to the export table, although it's a little less convoluted. There are three main structures used when importing a symbol: the *import directory table* (IDT as shown in Figure 3.7), *import lookup table* (ILT as shown in Figure 3.8), and the *hint/name table*. The IDT contains a few fields; the ones discussed here are the *ILT RVA*, the *name RVA*, and the *IAT RVA*. All of these are fairly self-explanatory except for the name RVA, which is also simple enough in that it is the RVA of the null-terminated ASCII string of the name of the DLL to be imported. The ILT and IAT are arrays of 32-bit integers (on PE32), with each entry being a bit-field. The high-order bit of an entry indicates whether the import is done by name or by ordinal; if the bit is set it is imported by ordinal. If the import is done by ordinal, bits 0 to 15 represent the ordinal to import. If it's being imported by name, then bits 0 to 30 represent a 31-bit RVA into the hints/name table for the name of the import. The hints/name table is yet another fairly simple table. The first two bytes of each entry serve as a "hint" to the loader.

Figure 3.7 IDT

Offset	Size	Field Name
0	4	**Import Lookup Table RVA**
4	4	Time/Date Stamp
8	4	Forwarder Chain
12	4	**Name RVA**
16	4	**Import Address Table RVA**

Bold indicates discussed/referenced field

Figure 3.8 ILT

Bits	Size	Field Name
31	1	**Ordinal/Name Flag**
15-0	16	**Ordinal Number**
30-0	31	**Hint/Name Table RVA**

Bold indicates discussed/referenced field

This hint is used as an index into the export name pointer table in the target DLL. If the entry matches, then this is used; otherwise a search for the name is performed. The next element is a variable length null-terminated ASCII string that is the name of the function to import, potentially followed by a trailing null padding byte, in order to have the next entry properly aligned.

NOTE

An interesting side note is that, while the IAT and ILT are supposed to contain the same data until the symbols are actually bound, the author has found that this was not always the case. In the distant past, while writing a tool to parse the PE format, it was found that some compilers would move the ILT into the .text segment and its contents would actually be different from expected! It was also found that some compilers didn't make use of the ILT at all and instead only used the IAT. When manually parsing the format, be aware of subtle nuances like this. Both Microsoft and Borland like to take shortcuts when possible.

Finally, we move into the *Load Configuration* structure, as shown in Figure 3.9. This structure was supposedly used in limited cases in Windows NT in a very different manner from how it is used in the post-Windows 2000 world. In Windows XP and later, this section is used by SafeSEH to register valid exception handlers with the system, thus avoiding the issue of an attacker overwriting an SEH entry and causing an exception to be raised and thus having their code executed. If the *IMAGE_DLLCHARACTERISTICS_NO_SEH* field is not set in the *DLL Characteristics* field of the optional header, and an exception handler is not in this list when the system is attempting to call it, then the process is aborted. In the *Load Configuration* structure, there are only three fields we would find interesting: the *security cookie*, *structured exception (SE) handler table*, and *structured exception handler count*. The *security cookie* is not actually the cookie itself, but rather a pointer to it. This cookie is used in a number of ways, most notoriously when the */GS* flags are specified to the Microsoft compiler, which implements stack cookies to prevent stack-based buffer overflows. The *SE handler table* is a sorted table of RVAs that correspond to valid SEH handlers for that particular image. The *SE handler count* is a count of the total number of handlers.

Figure 3.9 Load Configuration Structure

Offset	Size	Field Name
0	4	Characteristics
4	4	TimeDateStamp
8	2	MajorVersion
10	2	MinorVersion
12	4	GlobalFlagsClear
16	4	GlobalFlagsSet
20	4	CriticalSectionDefaultTimeout
24	8	DeCommitFreeBlockThreshold
32	8	DeCommitTotalFreeThreshold
40	8	LockPrefixTable
48	8	MaximumAllocationSize
56	8	VirtualMemoryThreshold
64	8	ProcessAffinityMask
72	4	ProcessHeapFlags
76	2	CSDVersion
78	2	Reserved
80	8	EditList
88	4	**SecurityCookie**
96	4	**SEHandlerTable**
104	4	**SEHandlerCount**

Bold indicates discussed/referenced field

Executable and Linking Format

The Executable and Linking Format (ELF) was the result of work done at Unix System Laboratories and was eventually published as part of the System V Application Binary Interface (ABI) and then later adopted in Tool Interface Standard. Interestingly enough, the original name for the format was Extensible Linking Format, most likely a result of many prior file formats not supporting the dynamic linking of external libraries. Since its official adoption as the file format of choice for Unix and Unix-like operating systems, it has become the de facto standard across the board in the Unix world, with nearly every vendor either using it as their native format or supporting it through a thin abstraction layer. It is used in everything from Linux, Solaris, IRIX and the BSDs to the Playstation. Therefore, unless the world in which you operate is entirely Windows based, you will encounter ELF files in pretty short order.

In the ELF header (shown in Figure 3.10), there are no entirely strict sizes; everything is defined relative to the native sizes of the processor, and a lot of processors use it. For this reason, this chapter only references IA32. The ELF header and indeed the format are a lot more straightforward and interesting to us as reverse engineers, as you might ascertain from the number of elements in the header that are touched upon in this chapter. The *e_ident* field identifies the ELF file: it identifies the file as an ELF file, identifies the native word size of the processor, specifies the intended byte-ordering, and finally the version of the ELF header. The field is an array of unsigned characters, 16 in total. The first four bytes are the "magic" field with the values of 0x7F, E, L and F. The next byte specifies the word size or class, with a value of 0 indicating that it is an invalid class, a value of 1 indicating that it is 32-bit, and a value of 2 indicating that it is 64-bit. The next byte specifies the encoding of data within the file, with a possible value of 0, 1 or 2. Zero again signifies an invalid value, 1 indicates that the data is encoded in two's complement values with the least significant byte occupying the lowest address, and 2 indicating that the encoded values are encoded in two's complement values with the most significant byte at the lowest address. The next byte in the *e_ident* field specifies the version of the ELF header and, as we will see shortly, is somewhat redundant. It should be set to 1, indicating that it is using the current version of the ELF specification. Finally, the rest of the bytes in the array are currently unused and reserved. They currently serve as padding and the specification suggests that programs parsing the header ignore any values in the field. Following the *e_ident* field is the *e_type* field, which identifies what type of executable the file is. The possible values are 0 indicating that there is no file type, 1 indicating that the file is a relocatable file, 2 indicating that it is an executable file, 3 that it is a shared object file, and 0xFF00 and 0xFFFF are marked as being processor specific.

Figure 3.10 ELF Header

Offset	Size	Field Name
0	16	**e_ident**
16	2	**e_type**
18	4	**e_machine**
22	4	**e_version**
26	4	**e_entry**
30	4	**e_phoff**
34	4	**e_shoff**
38	4	**e_flags**
42	2	**e_ehsize**
44	2	**e_phentsize**
46	2	**e_phnum**
48	2	**e_shentsize**
50	2	**e_shnum**
52	2	**e_shstrndx**

Bold indicates discussed/referenced field

After the *e_type* field, we have the *e_machine* field, which indicates the type of processor the file was built for. The only value of substance for us is the value 3, which indicates that it is for an IA32 machine; the other values are for more esoteric architectures, such as SPARC, Motorola, MIPS and IA 8086. *e_version* has the same values as the version in the *e_ident* header, with 0 indicating that the version is invalid and 1 indicating that the version is the current version. The *e_entry* member sounds exactly like what it is—it holds the VA of the entry point of the application if applicable; otherwise the field is set to 0. The *e_phoff* field yields the file offset in bytes of the file's program header table. The *e_shoff* field yields the file offset in bytes of the file's section header table; both of these fields *e_phoff* and *e_shoff* are only present if the file has the table. Otherwise they're initialized to 0. The *e_flags* field contains processor-specific flags. However, the IA32 architecture specifies no flags and therefore the field will be (should be) 0. The *e_ehsize* field holds the size of the ELF header, the *e_phentsize* and *e_shentsize* indicate the size in bytes of one entry in the program header table and section header table, respectively. The *e_phnum* and *e_shnum* fields indicate the number of entries in the program header table and the section header table. Thus, to calculate the size of the program header table, you would multiply the *e_phentsize* and the *e_phnum* fields. Once again, if any of these tables do not exist, the field's values are initialized to 0. Finally, at the end of the ELF header we have the *e_shstrndx* field, which holds the index for the section name string table in the section header table, if applicable.

The ELF section header table is an array of section header structures. Each structure is of the format displayed in Figure 3.9 and is again fairly straightforward. That said, in the array there are certain indexes that hold special values; these special indexes are shown in Table 3.3.

Table 3.3 Special Indexes for ELF Section Header Table Array

Name	Index/Value
SHN_UNDEF	0
SHN_LORESERVE	0xFF00
SHN_LOPROC	0xFF00
SHN_HIPROC	0xFF1F
SHN_ABS	0xFFF1
SHN_COMMON	0xFFF2
SHN_HIRESERVE	0xFFFF

SHN_UNDEF marks an undefined, absent or generally meaningless section reference. It is important to note that, although undefined, the section header table (if present) always contains an SHN_UNDEF entry at index zero; thus, if the *e_shnum* field states that there are ten fields, there are actually nine plus the SHN_UNDEF entry. SHN_LORESERVE specifies the lower bound of reserved index ranges. SHN_LOPROC and SHN_HIPROC specify the range of entries reserved for processor-specific entries. SHN_ABS specifies absolute values for the relevant symbols. This essentially means symbols referenced are not affected by relocation. SHN_COMMON is for common symbols such as unallocated external variables in C, and finally SHN_HIRESERVE specifies the upper bound of reserved index ranges. Every section in an ELF file has exactly one section header describing it; the sections described are contiguous although potentially empty and the sections themselves may not overlap. The elements of a section header are as listed in Figure 3.11.

Figure 3.11 Section Header

Offset	Size	Field Name
0	2	**sh_name**
2	2	**sh_type**
4	2	**sh_flags**
6	4	**sh_addr**
10	4	**sh_offset**
14	2	**sh_size**
16	2	**sh_link**
18	2	**sh_info**
20	2	sh_addralign
22	2	**sh_entsize**

Bold indicates discussed/referenced field

The *sh_name* element is an index into the section header string table, which specifies the name of the section. The *sh_type* element determines the type of section contents and semantics; specifically the following types are defined in Table 3.4.

Table 3.4 Types of Section Contents and Semantics

Name	Value
SHT_NULL	0
SHT_PROGBITS	1
SHT_SYMTAB	2
SHT_STRTAB	3

Continued

Table 3.4 Continued

Name	Value
SHT_RELA	4
SHT_HASH	5
SHT_DYNAMIC	6
SHT_NOTE	7
SHT_NOBITS	8
SHT_REL	9
SHT_SHLIB	10
SHT_DYNSYM	11
SHT_LOPROC	0x70000000
SHT_HIPROC	0x7FFFFFFF
SHT_LOUSER	0x80000000
SHT_HIUSER	0xFFFFFFFF

The SHT_NULL value indicates that the section header entry is inactive and does not have an associated section. The SHT_PROGBITS value indicates that the section holds data defined by the program itself and that the format is only known to the program. The SHT_SYMTAB and SHT_DYNSYM sections define symbol tables. An application may have exactly one SHT_SYMTAB section (although likely it will have none). It typically contains a complete table of symbols, whereas the SHT_DYNSYM section defines a minimal set of symbols to be used for dynamic linking. Neither of these sections is gone into in detail since it is unlikely you will be reversing a file that actually has symbols—however, the interested reader is encouraged to read the ELF specification. The SHT_STRTAB is a section that holds a string table; a file can have more than one string table section and these sections are explained in more detail later on in this chapter. SHT_RELA sections contain relocation entries with explicit addends. SHT_HASH sections contain a symbol hash table and at the moment only one section of this type is allowed per object. This section is required by all objects participating in dynamic linking. The SHT_DYNAMIC type is used for dynamic linking. SHT_NOTE indicates that the section holds information that marks the file in some way. This section is not described for brevity's sake. SHT_NOBITS indicates that the section occupies no space in the file but otherwise resembles an SHT_PROGBITS section; the most well-known SHT_PROGBITS section is the .BSS. SHT_REL holds relocation entries without explicit addends, and the SHT_SHLIB section is reserved but has unspecified semantics. SHT_LOPROC and SHT_HIPROC define a range of sections that are reserved for processor-specific semantics, whereas SHT_LOUSER and SHT_HIUSER are the same except they are reserved for the application.

Following the *sh_type* field we have the *sh_flags* field, which defines various 1-bit attributes for the section. The attributes are enumerated in Table 3.5.

Table 3.5 Attributes

Name	Value
SHF_WRITE	0x1
SHF_ALLOC	0x2
SHF_EXECINSTR	0x4
SHF_MASKPROC	0xF0000000

The SHF_WRITE flag indicates that the section will be writeable. SHF_ALLOC indicates that the section should actually reside in memory at run-time. SHF_EXECINSTR indicates that the section contains executable instructions. Finally, SHF_MASPROC indicates that the section is reserved for processor-specific uses.

Next we have the *sh_addr* field; it provides the preferred VA at which the section should start. The *sh_offset* is similar, except that it provides a file offset from the beginning of the file to the beginning of the section. The *sh_size* field is the size of the section, unless it is of type SHT_NOBITS, although a SHT_NOBITS section can have a nonzero *sh_size*; it just does not occupy space in the physical file. The *sh_link* and *sh_info* members are related in that both values are subject to interpretation dependent on the section type. In the case of a SHT_DYNAMIC section, the *sh_link* member indicates the section header table index of the string table used by the section, and has a *sh_info* field of zero. For SHT_HASH the *sh_link* member holds the section header table index of the symbol table that applies to the hash table, and again has an *sh_info* value of zero. For a full list of values, please refer to the ELF specification. Finally, the *sh_entsize* member applies to certain sections that have fixed-size tables inside of them; for instance, given a section with a symbol table, this entry would indicate the length of the symbol table.

Now that we have some comprehension of the various sections, Table 3.6 denotes sections that are fairly common and standard, with their types, attributes and a brief description. It should be noted that sections whose name has a leading period (such as .bss) are reserved for use by the system.

Table 3.6 Common Sections

Section Name	Type	Attributes	Description
.bss	SHT_NOBITS	SHF_ALLOC SHF_WRITE	Holds uninitialized data, is initialized with zeros at load time; typically globally scoped
.comment	SHT_PROGBITS	n/a	Contains version control information
.data	SHT_PROGBITS	SHF_ALLOC SHF_WRITE	Contains initialized data that contributes to applications memory image, typically globally scoped
.data1	SHT_PROGBITS	SHF_ALLOC, SHF_WRITE	Contains initialized data that contributes to applications memory image, typically globally scoped
.debug	SHT_PROGBITS	n/a	Unspecified contents used for symbolic debugging
.dynamic	SHT_DYNAMIC	SHF_ALLOC SHF_WRITE (processor specific)	Contains dynamic linking information, whether SHF_WRITE is specified or not is processor specific
.dynstr	SHT_STRTAB	SHF_ALLOC	Contains strings needed for dynamic linking
.dynsym	SHT_DYNSYM	SHF_ALLOC	Contains strings needed for dynamic linking
.fini	SHT_PROGBITS	SHF_ALLOC SHF_EXECINSTR	Contains executable instructions used in application termination, such as destructors
.got	SHT_PROGBITS		Described in detail later
.hash	SHT_HASH	SHF_ALLOC	Contains a symbol hash table
.init	SHT_PROGBITS	SHF_ALLOC SHF_EXECINSTR	The inverse of .fini, for ex., contains constructors

Table 3.6 Continued

Section Name	Type	Attributes	Description
.interp	SHT_PROGBITS	SHF_ALLOC	Holds the path name of a program interpreter; if the file contains a loadable segment then the SHF_ALLOC attribute will be set
.line	SHT_PROGBITS	n/a	Contains line information for debugging
.note	SHT_NOTE	n/a	Can be used by the implementation to allow stigmatic marking of an executable
.plt	SHT_PROGBITS		Described in detail later
.rel<name>	SHT_REL	SHF_ALLOC	Contains relocation information; if there is a loadable segment then SHF_ALLOC will be set. Traditionally in the place of <name> is the name of the section that the relocations are for, such as .rel.text
.rela<name>	SHT_RELA	SHF_ALLOC	Contains relocation information; if there is a loadable segment then SHF_ALLOC will be set. Traditionally in the place of <name> is the name of the section that the relocations are for, such as .rela.text

Continued

Table 3.6 Continued

Section Name	Type	Attributes	Description
.rodata	SHT_PROGBITS	SHF_ALLOC	Contains read-only data, such as constant strings
.rodata1	SHT_PROGBITS	SHF_ALLOC	Contains read-only data, such as constant strings
.shstrtab	SHT_STRTAB	n/a	Contains section names
.strtab	SHT_STRTAB	SHF_ALLOC	Contains strings, typically names associated with symbol table entries. If there is a loadable section then SHF_ALLOC will be specified.
.symtab	SHT_SYMTAB	SHF_ALLOC	Contains a symbol table (not described in this chapter). If there is a loadable section, then SHF_ALLOC will be specified
.text	SHT_PROGBITS	SHF_ALLOC SHF_EXECINSTR	The executable instructions that make up the program

The *program header table* is an array of *program header* structures; these structures define segments and generally define how to load the binary to the operating system (OS). The size of the table and the number of entries are specified in the ELF header. Some of the segments are supplementary, whereas others contribute to the process image. Just like everything else in the ELF (sans the ELF header itself) file, there is no specific ordering of the segments nor specific offset to the program header table; this is defined solely by the ELF header. In Figure 3.12 to the right you will find a diagram detailing the ordering and members of an *Elf32_Phdr* structure. The *p_type* field indicates what type of segment is being described and by implication tells the system how to interpret its contents. The defined values are shown in Table 3.7.

Table 3.7 Defined Values

Name	Value
PT_NULL	0
PT_LOAD	1
PT_DYNAMIC	2
PT_INTERP	3
PT_NOTE	4
PT_SHLIB	5
PT_PHDR	6
PT_LOPROC	0x700000000
PT_HIPROC	0x7FFFFFFF

Figure 3.12 Program Header Structure

Offset	Size	Field Name
0	4	**p_type**
4	4	**p_offset**
8	4	**p_vaddr**
12	4	p_paddr
16	4	**p_filesz**
20	4	**p_memsz**
24	4	**p_flags**
32	4	p_align

Bold indicates discussed/referenced field

Segments of type PT_NULL are unused; the values of its other members are undefined (and thus should be ignored). The reasoning behind having another NULL type of segment is to allow segments to be defined but ignored by the implementation. Segments of type

PT_LOAD are actually loaded into memory preferably at the address *p_vaddr*. The first *p_filesz* bytes at *p_offset* from the file's memory-mapped base are loaded into memory. If *p_memsz* is larger than *p_filesz* then these bytes are also mapped into the segment and zero filled; it is invalid for a *p_filesz* member to be greater in value than *p_memsz*. The PT_INTERP segment, if present, must precede any PT_LOAD segments as it indicates the path name of the program interpreter. This segment can occur only once in the file (if the file wishes to be valid). Segments of type PT_DYNAMIC are related to dynamic linking. PT_NOTE is relatively unimportant but allows interacting applications to check conformance (that is, GLIBC version). PT_SHLIB is defined but reserved, and files containing a PT_SHLIB segment do not conform to the ABI. The PT_PHDR segment specifies the size and location of the program header table. This specification applies to both in the physical file and in the memory image. Like the PT_INTERP segment, it can only occur once and if present must occur before any PT_LOAD segments. Finally, PT_LOPROC and PT_HIPROC are reserved ranges for processor-specific functionality.

As one might have guessed from the previous description, the *p_offset* member specifies the offset of the segment from the beginning of the file. The *p_vaddr* member specifies the preferred VA of the segment. The *p_filesz* and *p_memsz* elements specify the size of the segment in the physical file and in memory, respectively, and finally the *p_flags* specifies attributes of the segment. The three possible values are PF_R, PF_W and PF_X for read, write and execute, respectively.

Now that we have some basic understanding of segments, it's possible to talk a bit about the differences between executable images, shared library images, or images that have *address space layout randomization* (ASLR) applied. Typically, in order to load an executable image, the address for each segment used when building the image must be included. This address is specified in the segment's *p_vaddr* member. This is the result of the image having absolute references that would break if the addresses were changed. ASLR images and shared library images typically get around this restriction by using what is known as *position independent code* (PIC). The general idea behind PIC is that, instead of using absolute references to some piece of data, relative references are used. For instance, whereas in traditional code you may access a variable at address XYZ, in PIC you would reference that by another means—that is, relative to your current position. In the case of an application using a shared library to access commonly used functions, such as is common with the standard C library, a series of intermediaries are used—in the case of ELF, it is the *global offset table* (GOT or .got), the *dynamic segment/section* (_DYNAMIC or .dynamic) and the *procedure linkage table* (PLT or .plt). These three segments are integrally interrelated and make up one of the major reasons for adopting ELF over older standards such as a.out—namely, standards-supported dynamic linking. The .dynamic segment is present in every executable image that takes part in dynamic linking; this segment is referenced by the symbol _DYNAMIC, which is an array of structures as illustrated in Figure 3.13.

Figure 3.13 Dynamic Structure

Offset	Size	Field Name
0	4	**d_tag**
4	4	**d_val**
4	4	**d_ptr**

Bold indicates discussed/referenced field

The dynamic structure contains two values, a tag followed by a union. The tag determines how the union will be interpreted. The *d_val* member contains an integer with various interpretations that are described below, whereas the *d_ptr* member contains a VA. As you well know, the compile-time VA and the runtime VA might differ and the relocations section does not contain relocations for the _DYNAMIC array. In Table 3.8 you will find several defined *d_tag* types, and whether they are optional or mandatory. This is not the full list but only what is relevant to us, and the interested reader is again highly encouraged to refer back to the ELF specification for a more detailed and complete explanation.

Table 3.8 Defined *d_tag* Types

Name	Value	d_val or d_ptr?	executable	Shared object
DT_NULL	0	ignored	mandatory	mandatory
DT_PLTRELSZ	2	d_val	optional	optional
DT_PLTGOT	3	d_ptr	optional	optional
DT_FINI	13	d_ptr	optional	optional
DT_PLTREL	20	d_val	optional	optional
DT_JMPREL	23	d_ptr	optional	optional

The DT_NULL element marks the end of the _DYNAMIC array, and thus it is a necessary element. Aside from this element, there is no inherent ordering within the array. The DT_PLTRELSZ element holds the total size of relocation entries associated with the PLT. If a DT_JMPREL element is present, then a corresponding DT_PLTRELSZ entry must also be present. DT_PLTGOT entries hold an address associated with either the GOT or the

PLT, both of which are described in further detail below. DT_FINI elements hold the address of a termination function, or a destructor, which potentially would be useful to hackers and therefore is of interest to us and is covered briefly later on. Finally, the DT_JMPREL entry contains a pointer to relocation entries associated only with the PLT. Separating these relocations allows the linker to ignore them during image initialization and use a form of linking known as *lazy binding*. Quite simply, lazy binding defers the relocation until the actual use of that symbol. For instance, consider the following C program:

```
#include <stdio.h>
#include <stdlib.h>
int
main(void)
{
        unsigned int cnt;
        for (cnt = 0; cnt < 2; cnt++)
          printf("cnt: %u\n", cnt);
        exit(EXIT_SUCCESS);
}
```

In this application, we have two (visible) standard library functions called *printf()* and *exit()*. In a lazy binding scenario, neither symbol is resolved until the last possible minute. The first time *printf()* is called the symbol is resolved, incurring the overhead of relocation then, instead of at initialization. However, the second time *printf()* is called, the symbol has already been resolved and the overhead is not incurred to resolve *printf()* again. However, that overhead is again incurred to resolve *exit()*. The advantage of this, of course, is the increase in speed and efficiency; after all, not every symbol is going to be resolved. Furthermore, it makes dynamic module loading easier to cope with. However, the downside of this comes in the form of attack surface—it makes a buggy program more easily exploited. In recent years, a flag passed to hardened GCC tool-chains called *relro* does all relocations at program initialization and then disallows writing to the segments at runtime. This means that a potential attacker cannot take advantage of some dubious pointer arithmetic or a buffer overflow and write into these sections and then return execution flow back into these sections. Typical sections to be marked this way are .init/.ctors, .fini/.dtors, the PLT, GOT and .dynamic, although specifics depends upon architecture. This methodology is becoming more and more common, and as a reverse engineer it is likely that as time progresses the chances of your running across lazy binding decreases.

The global offset table, which on IA32 platforms is accessible under the symbol _GLOBAL_OFFSET_TABLE, is an array of addresses. These addresses are absolute references and allow the position-independent code to have relative references. Thus, the PIC code will obtain the address of the GOT and extract absolute references from its relative ones. The symbol _GLOBAL_OFFSET_TABLE need not refer to the beginning of the .got segment, and thus negative and positive indexes are potentially valid. When the image is loaded, the dynamic linker walks through the relocations and looks for entries of a specific type and replaces their entries in the GOT with their absolute addresses, effectively getting around the limitations of the static linker. The first element of the GOT is a special entry that contains the address of the _DYNAMIC structure; this allows the dynamic linker to process the GOT by finding itself in the _DYNAMIC structure without having to depend on any relocations. Furthermore, on IA32 the second and third entries are also reserved to have special values. The GOT redirects position-independent addresses to absolute locations, whereas the PLT does the same but for functions. The PLT determines the function's absolute address and updates the GOT as necessary. The exact implementation of the PLT varies, depending on whether it was compiled PIC or not. A non-PIC entry looks something like the following:

```
PLT
.PLT0
       push   address_of_GOT+0x04
       jmp    [address_of_GOT+0x08]
       nop
       nop
       nop
       nop
.PLT1:
       jmp    [name1_in_GOT]
       push   offset
       jmp    [.PLT0+$]
.PLT2:
       jmp    [name2_in_GOT]
       push   offset
       jmp    [.PLT0+$]
       ...
```

whereas a PIC PLT might look something like this:

```
PLT
  .PLT0
       push  [ebx+0x04]
       jmp   [ebx+0x08]
       nop
       nop
       nop
       nop
```

```
.PLT1
        jmp     [ebx+name1]
        push    offset
        jmp     [.PLT0+$]
.PLT2
        jmp     [ebx+name2]
        push    offset
        jmp     [.PLT0+$]
```

With all this taken into context, in order to resolve dynamic references, the linker and the application work in tandem according to the following steps:

0. Upon creation of the image, the second and third entries in the GOT are set to their special values as defined below.
1. If the PLT is PIC, then the address of the GOT must reside in the *ebx* register. The calling function is responsible for placing the address into this register.
2. Assume that the application is trying to call *name1* which can be found in the label .PLT1.
3. The first instruction under that label is a jump into the GOT, which initially contains the address of the *push* and *jmp* instructions following the *jump* instruction into the GOT.
4. The application then pushes the address of the relocation entry, represented in this case by the variable named *offset*. This offset will specify GOT entry used in the prior jump along with a symbol table index, *name1* in this instance.
5. The application then jumps to .PLT0 and pushes the address of the second element of the GOT onto the stack, giving the dynamic linker a word to reference for identification purposes, and then transfers control to the third GOT entry, which hands control to the dynamic linker.
6. The dynamic linker unwinds the stack and retrieves the identifying information, finds the absolute address for the symbol and stores it in the related GOT entry, and then hands control to the requested function.
7. Further calls to this function will skip the push of *offset* and will jump to .PLT0 as a result of having the GOT entry modified.

So, as you can see, dynamic linking is accomplished by indirection and abstraction. The application doesn't know beforehand exactly what address it's calling, and in turn calls into the PLT, which in turn jumps to the GOT; if the address has not already been resolved, control is handed back into the PLT, which pushes the relocation entry and jumps to the first entry in the PLT, which then hands control to the dynamic linker.

NOTE

As previously mentioned, several advances have been made in the not-so-distant past that require that lazy binding be turned off so that relocations can occur at initialization instead of at runtime. If you think hard enough, now that you know how the GOT/PLT works, you might realize why. If as an attacker I can overwrite a GOT entry, then it's really only a matter of the application calling that function again before my shellcode obtains control. This technique has been documented in several places; one of the white papers can be found at the following URL: www.milw0rm.com/papers/3.

Similarly, a given image destructor can be attacked by overwriting data in .dtors, which typically contains the addresses of functions in .fini. By overwriting an address there, it becomes potentially possible for the application to have a rogue function called upon program execution. This technique was documented by Juan M. Bello Rivas and his white paper can be found at the following URL: http://synnergy.net/downloads/papers/dtors.txt

Summary

In conclusion, we've taken you on a brief tour of the PE and ELF file formats, giving you a basic understanding of these formats and hopefully giving you the knowledge needed to perform limited manual analysis of either one. You should be able to determine the imports and exports of a PE (and consequently have a basic understanding for rebuilding the imports section in a packed PE), and have a decent understanding of how dynamic linking occurs in ELF files. In both formats, you should be fairly comfortable with their members and structure and generally have some understanding of how a linker and loader operate on either of the formats. As suggested throughout the chapter, the reader is encouraged to read the original specifications themselves, as this chapter is far from complete and tries to emphasize only elements that the author thought would be most important. Some concepts that you might find particularly interesting or useful may not be covered. The ELF specification can be found at the following URL: www.muppetlabs.com/~breadbox/software/ELF.txt or via your favorite search engine by searching for "ELF specification." The PE specification can be obtained from Microsoft's website at www.microsoft.com/whdc/system/platform/firmware/PECOFF.mspx.

Chapter 4

Walkthroughs One and Two

Solutions in this chapter:

- Understanding Execution Flow
- Tracing Functions
- Recovering Hard Coded Password
- Finding Vulnerable Functions
- Backtracing Execution
- Crafting a Buffer Overflow

☑ Solutions Fast Track

☑ Frequently Asked Questions

Introduction

In this chapter we will step away from the basics of IDA and dive straight into applying our knowledge. This is a good starting point for the average computer or security professional with a general knowledge of security, assembly basics and programming. We will begin by figuring out exactly what our first example binary does, and then move to applying this knowledge in common practices within the security industry. Specifically, we'll see if we can find the password it's asking for when it first starts and then leverage this knowledge in order to find vulnerability within the binary. Applying these two approaches, we'll finally be able to understand the steps needed to actually exploit the application.

The example code and binaries we will be using for this chapter are available for download from the Syngress website. The download file is called StaticPasswordOverflow.zip.

Following Execution Flow

The first step in reversing any binary on the planet is determining exactly what it is doing and how it is doing it. Let's jump into the immediate task of following the instructions of our application step by step, and take notes on the general operations within the binary. To begin, let's go straight ahead into the first useful chunk of code. Although it's personal preference, I prefer using a notepad or notebook of some sort so I can keep my thoughts as I move along, write down addresses, and generally keep track of everything. You never know when you might need a note from the beginning of your reversing, and typing numbers has, for me, always been much slower than writing them down. Plus, it's much easier to draw pictures on paper!

```
.text:00401270 ; int __cdecl main(int argc,const char **argv,const char *envp)
.text:00401270 Dst    = byte ptr -80h
.text:00401270 argc   = dword ptr 8
.text:00401270 argv   = dword ptr 0Ch
.text:00401270 envp   = dword ptr 10h
.text:00401270        push    ebp
.text:00401271        mov     ebp, esp
.text:00401273        sub     esp, 80h
.text:00401279        push    offset aReverseEnginee
.text:0040127E        call    sub_401554
.text:00401283        add     esp, 4
.text:00401286        push    offset aPleaseProvideT
.text:0040128B        call    sub_401554
.text:00401290        add     esp, 4
.text:00401293        push    80h    ; Size
.text:00401298        push    0      ; Val
```

```
.text:0040129A       lea     eax, [ebp+Dst]
.text:0040129D       push    eax    ; Dst
.text:0040129E       call    _memset
.text:004012A3       add     esp, 0Ch
.text:004012A6       lea     ecx, [ebp+Dst]
.text:004012A9       push    ecx
.text:004012AA       push    offset a127s    ; "%127s"
.text:004012AF       call    _scanf
.text:004012B4       add     esp, 8
.text:004012B7       lea     edx, [ebp+Dst]
.text:004012BA       push    edx    ; Str2
.text:004012BB       call    sub_4011C0
.text:004012C0       add     esp, 4
.text:004012C3       movsx   eax, al
.text:004012C6       test    eax, eax
.text:004012C8       jge     short loc_4012D9
.text:004012CA       push    offset aYouFailed_
.text:004012CF       call    sub_401554
.text:004012D4       add     esp, 4
.text:004012D7       jmp     short loc_4012E6
.text:004012D9 loc_4012D9: ; CODE XREF: _main+58
.text:004012D9       push    offset aYouWon_Goodbye
.text:004012DE       call    sub_401554
.text:004012E3       add     esp, 4
..text:004012E6 loc_4012E6: ; CODE XREF: _main+67
.text:004012E6       mov     eax, 1
.text:004012EB       mov     esp, ebp
.text:004012ED       pop     ebp
.text:004012EE       retn
.text:004012EE _main endp
```

At a glance, we can see that the *main* function doesn't really do much of anything. It has a few calls and a few conditional statements. Also, just from the strings within some of these statements, it looks like we can assume there is a success/failure statement within this code; thus, the strings containing "YouFailed" and "YouWon". We could switch over to the graph view right away to determine how these conditionals work, but first we will get an understanding for how this entire function works, so we have no surprises later on.

```
.text:00401279       push    offset aReverseEnginee
.text:0040127E       call    sub_401554
.text:00401283       add     esp, 4
```

```
.text:00401286        push    offset aPleaseProvideT
.text:0040128B        call    sub_401554
```

Here we can see that, after setting up the stack, it's pushing some static strings into the buffer and calling a function. By looking at the strings, it's safe to assume this is some sort of startup header printing. However, it looks like IDA cannot determine exactly what function this binary is calling. Let's go ahead and follow the call and see where it's going; select the call instruction and press **Enter** to jump to that location.

```
.text:00401554 ; int printf(const char *,…)
.text:00401554_printf        proc near ; CODE XREF: sub_401000+65
.text:00401554; sub_401000+C0
```

Whoops! It looks like IDA didn't want to identify what exactly this function was; it's just a statically compiled *printf*. It's fairly safe to assume this function isn't doing anything odd or funky right now, so we'll chalk that one up as a simple print function and get on with it. Let's press the **Backspace** key in order to get back to our entry point and continue.

NOTE

Some functions may not always be what they appear to be; always give such obviously named static functions a good look before assuming the name is real. You never know what the bad guys might be trying to hide with clever names.

Since we know what that routine is, we should quickly rename it within IDA so we don't have to worry about it confusing us later. The easiest way to accomplish this is to simply click the name of the instruction, and press the **N** key, which will pop up the rename dialog box. For the sake of ease, we'll rename this function *printf*, since that is what it really is. Now that we have that out of the way and have confirmed that it is just printing strings as a sort of startup process, we'll continue down the code.

```
.text:00401290        add     esp, 4
.text:00401293        push    80h      ; Size
.text:00401298        push    0        ; Val
.text:0040129A        lea     eax, [ebp+Dst]
.text:0040129D        push    eax      ; Dst
.text:0040129E        call    _memset
.text:004012A3        add     esp, 0Ch
.text:004012A6        lea     ecx, [ebp+Dst]
```

```
.text:004012A9        push   ecx
.text:004012AA        push   offset a127s   ; "%127s"
.text:004012AF        call   _scanf
```

Stepping through this set of instructions, it seems obvious that it is calling *memset()* to fill a buffer, and then using *scanf()* in order to read into a buffer. Specifically, we can see that the *memset()* call is filling the first 0×80, or 128, bytes of the *Dst* stack buffer with 0×00, or NULL. This can be deduced by seeing the values being pushed prior to the call, where we see the following four instructions:

```
.text:00401293        push   80h     ; Size
.text:00401298        push   0       ; Val
.text:0040129A        lea    eax, [ebp+Dst]
.text:0040129D        push   eax     ; Dst
```

We can see here that the binary is pushing a size of 0×80, a value of 0 and finally pushing a pointer to the address of [ebp+Dst], our stack variable. Finally, we can also see that these same operations are being used for the *scanf()* call. Specifically, the instruction to load a pointer to the *Dst* buffer (lea ecx, [ebp+Dst]) is performed again, and the result of *scanf()* is saved within this buffer. We can also see that the *scanf()* call is correctly filling the buffer, with the *%127s* definition for its format string; thus only saving a maximum of 127 bytes to the buffer.

NOTE

If you don't feel comfortable enough with 16-base hex numbers yet, you can always right-click a numerical value within IDA and view or select a different base type for the numeral. Although IDA's default is hex values, you can click the value and press the **H** key in order to switch it to standard 10-base numbers.

Momentarily going back to the stack initialization portion of this function, we can check to make sure these sizes correspond with the actual size of the operations being performed on this variable. We can see IDA has already determined that this function did have a variable, and its size was 0×80 bytes long. Additionally, we can see the stack initialization calls performing this, thus confirming that this is the hard set size of this stack variable.

```
.text:00401270        Dst    = byte ptr -80h
......
.text:00401273        sub    esp, 80h
```

Now we have the uninteresting portions of the code out of the way and we understand what it all does. We can finally move forward to the interesting conditionals we saw within the code, which seem to be where all the magic must be happening. If you switch over to the graphing view now (press the **Spacebar**) you can see the conditional jumps that occur right after the *sprintf()* call, as shown in Figure 4.1. Additionally, you can see the mini-graph window, which becomes extremely useful with larger functions with many conditional jumps.

Figure 4.1 Graphing View

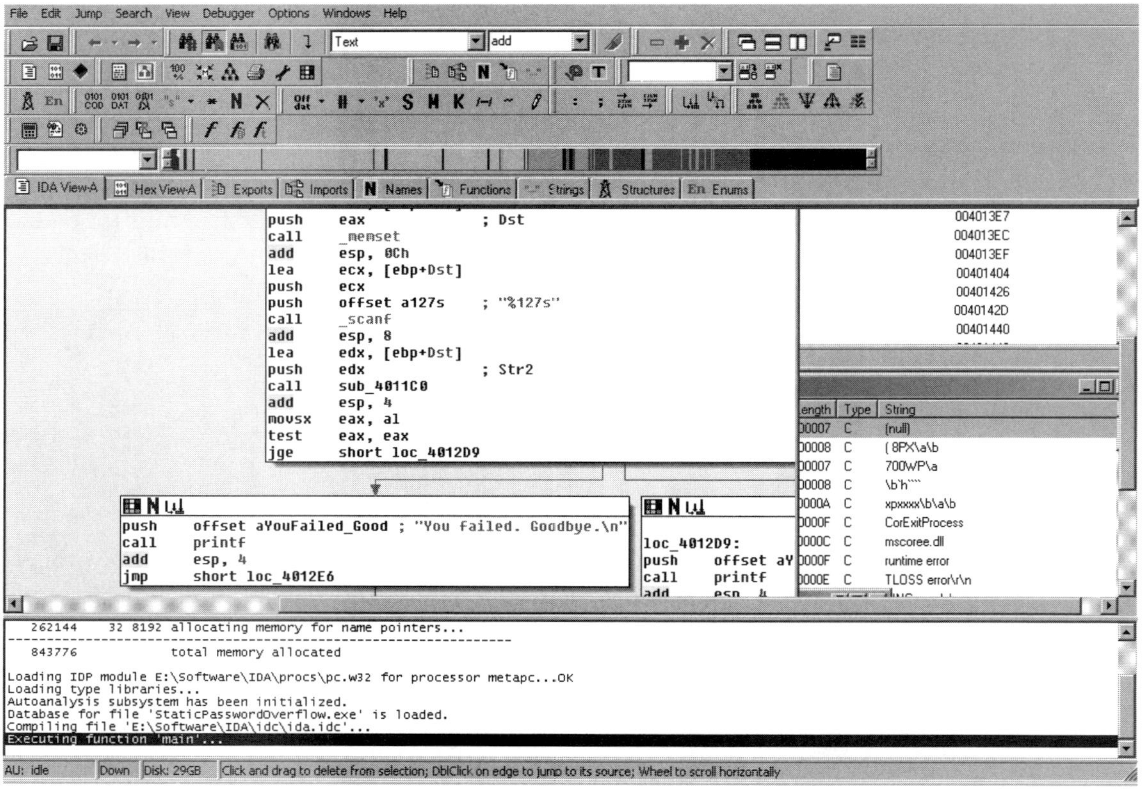

Reversing What the Binary Does

Moving past what seems to be the setup portion of this function, we can see that the function is calling a few subroutines prior to the conditional jump that has become our goal. Specifically, we can see that, prior to the conditional, the subroutine is calling another routine that is statically within the binary, and then immediately shifting the stack and performing the conditional.

Tools & Traps…

Binary Subroutines in IDA Pro

Be careful when analyzing subroutines within binaries that don't appear to be readily identifiable. Although it is commonly safe to assume that these are part of the actual executable itself, IDA Pro will identify many statically compiled libraries in this manner as well.

Always be sure to import debug symbols when possible and to label all functions once you have a good understanding of their purpose. ELF linux binaries are notorious for this, and many countless hours can be lost tracing statically compiled base libraries or GOT tables.

```
.text:004012A9        push   ecx
.text:004012AA        push   offset a127s  ; "%127s"
.text:004012AF        call   _scanf
.text:004012B4        add    esp, 8
.text:004012B7        lea    edx, [ebp+Dst]
.text:004012BA        push   edx    ; Str2
.text:004012BB        call   sub_4011C0
.text:004012C0        add    esp, 4
.text:004012C3        movsx  eax, al
.text:004012C6        test   eax, eax
.text:004012C8        jge    short loc_4012D9
```

We can move through the code immediately after the *scanf()* function call, which seems to handle the return value of *scanf()* and then prepare the stack to call this mysterious subroutine. Assuming you already have a good understanding of the stack structure and function call methods, let's move through this again as review. Two values are pushed just prior to the *scanf()* call, one of which is a static string within our binary. Using documentation available from many sources, we can easily deduce not only that these are the variables for *scanf()* being pushed, but also what they mean. This is extremely useful when reversing binaries that use less-common library functions. It will always be a generally good idea to look up library functions prior to attempting to reverse them; if they aren't critical to our goal, it can save lots of time to assume they perform as advertised. Below, you can see that *scanf()* is structured as such.

```
int scanf(const char *format, …);
```

Noting this, we can now see that what is actually being loaded and passed is first the pointer to a buffer, and then the format string (always remember, variables are pushed into the stack "backwards"). Next, the stack is shifted 8 bytes, and then the pointer to our buffer [ebp+Dst] is loaded into *edx* and then pushed onto the stack. After this, our magical subroutine is simply called. It will be safe to assume now that this function is performing some sort of mystical processing on the buffer being filled by *scanf()*, and returning an integer value which is then compared. This can be deduced by the instructions immediately after the call: movsx eax, al and test eax, eax. These instructions tell us that this portion of code is taking the return value of the called subroutine, call sub_4011C0, and conditionally jumping if it is greater than or equal to (the return value of calls are generally provided within the *eax* register).

So, as review, we now know that this mysterious subroutine is performing some sort of operation on our input value and providing the value that controls our conditional jump. We are getting closer to our goal! Now if we can discover what sort of operation this routine is performing and how to provide it with the correct value, we can control the conditional jump operation. For reference, let's label the call statement *input_process* and then switch our IDA Pro view to that function by double-clicking its name.

The Processing Subroutine

Now that we know this function is going to be what inevitably controls our success or failure within the application, let's step through the code in detail in order to understand what may be going on here. Additionally, looking at the code below, we will need to jump around a bit in order to truly understand what is happening. Not all reverse engineering can be performed by analyzing the executable from start to finish; analyzing end processing can sometimes yield a faster understanding of how everything got there.

```
.text:004011C0 ; int __cdecl input_process(char *Str2)
.text:004011C0 input_processproc near ; CODE XREF: _main+4B
.text:004011C0 Dst   = byte ptr -80h
.text:004011C0 var_7F= byte ptr -7Fh
.text:004011C0 var_7E= byte ptr -7Eh
.text:004011C0 var_7D= byte ptr -7Dh
.text:004011C0 var_7C= byte ptr -7Ch
.text:004011C0 var_7B= byte ptr -7Bh
.text:004011C0 var_7A= byte ptr -7Ah
.text:004011C0 var_79= byte ptr -79h
.text:004011C0 var_78= byte ptr -78h
.text:004011C0 var_77= byte ptr -77h
.text:004011C0 var_76= byte ptr -76h
.text:004011C0 var_75= byte ptr -75h
.text:004011C0 var_74= byte ptr -74h
.text:004011C0 var_73= byte ptr -73h
.text:004011C0 var_72= byte ptr -72h
.text:004011C0 var_71= byte ptr -71h
.text:004011C0 var_70= byte ptr -70h
```

```
.text:004011C0 Str2  = dword ptr 8
.text:004011C0       push   ebp
.text:004011C1       mov    ebp, esp
.text:004011C3       sub    esp, 80h
.text:004011C9       push   80h     ; Size
.text:004011CE       push   0       ; Val
.text:004011D0       lea    eax, [ebp+Dst]
.text:004011D3       push   eax     ; Dst
.text:004011D4       call   _memset
.text:004011D9       add    esp, 0Ch
.text:004011DC       mov    [ebp+var_70], 0
.text:004011E0       mov    [ebp+var_75], 73h
.text:004011E4       mov    [ebp+Dst], 74h
.text:004011E8       mov    [ebp+var_76], 73h
.text:004011EC       mov    [ebp+var_7F], 68h
.text:004011F0       mov    [ebp+var_7A], 6Dh
.text:004011F4       mov    [ebp+var_7C], 69h
.text:004011F8       mov    [ebp+var_7B], 73h
.text:004011FC       mov    [ebp+var_71], 64h
.text:00401200       mov    [ebp+var_74], 77h
.text:00401204       mov    [ebp+var_7E], 69h
.text:00401208       mov    [ebp+var_7D], 73h
.text:0040120C       mov    [ebp+var_78], 70h
.text:00401210       mov    [ebp+var_73], 6Fh
.text:00401214       mov    [ebp+var_72], 72h
.text:00401218       mov    [ebp+var_79], 79h
.text:0040121C       mov    [ebp+var_77], 61h
.text:00401220       mov    ecx, [ebp+Str2]
.text:00401223       push   ecx     ; Str2
.text:00401224       lea    edx, [ebp+Dst]
.text:00401227       push   edx     ; Str1
.text:00401228       call   _strcmp
.text:0040122D       add    esp, 8
.text:00401230       test   eax, eax
.text:00401232       jz     short loc_401247
.text:00401234       push   offset aInvalidPasswor ;
"\n******* INVALID PASSWORD *******\n"
.text:00401239       call   printf
.text:0040123E       add    esp, 4
.text:00401241       or     al, 0FFh
.text:00401243       jmp    short loc_40125D
```

```
.text:00401245        jmp    short loc_40125D
.text:00401247 loc_401247: ; CODE XREF: input_process+72
.text:00401247        mov    eax, [ebp+Str2]
.text:0040124A        push   eax
.text:0040124B        push   offset aSIsCorrect_ ; "%s is correct.\n\n"
.text:00401250        call   printf
.text:00401255        add    esp, 8
.text:00401258        call   sub_401000
.text:0040125D loc_40125D: ; CODE XREF: input_process+83
.text:0040125D; input_process+85
.text:0040125D        mov    esp, ebp
.text:0040125F        pop    ebp
.text:00401260        retn
.text:00401260 input_process       endp
```

We can see here that this function is a bit larger than the entrypoint, but it appears that IDA Pro has helped that to an extent. Briefly looking at the code, something immediately jumps out as an important portion of the function and as something we should begin to look at. At the address 00401228, there is a *strcmp()* call.

```
.text:00401220        mov    ecx, [ebp+Str2]
.text:00401223        push   ecx     ; Str2
.text:00401224        lea    edx, [ebp+Dst]
.text:00401227        push   edx     ; Str1
.text:00401228        call   _strcmp
```

Not only does it appear to be a *strcmp()* call to compare strings, but it is using the buffer that stores the input from *scanf()*, ebp+Dst. This looks promising for our goal: a string comparison against our input value, followed by a conditional statement that determines success or failure. But what can it be comparing against? Let's step back a bit in the code and take a look. In this string comparison, we can see the function is comparing the pointers in *ecx* and *edx*, which are respectively ebp+Str2 and ebp+Dst. Before actually determining what *Str2* is, let's rename it for reference to *StrPassword*.

Just prior to the comparison, there is a major chunk of *mov* instructions being performed on what appears to be a sequential segment of memory, which is actually within our preallocated buffers. A 128-byte buffer, or 0×80 bytes, is being allocated on the stack and being NULLed out by the *memset()* function.

```
.text:004011C0        push   ebp
.text:004011C1        mov    ebp, esp
.text:004011C3        sub    esp, 80h
.text:004011C9        push   80h     ; Size
.text:004011CE        push   0       ; Val
```

```
.text:004011D0        lea     eax, [ebp+Dst]
.text:004011D3        push    eax     ; Dst
.text:004011D4        call    _memset
```

After this buffer is prepared, it is sequentially filled with these static variables. Upon closer inspection, however, it seems all these values are relatively close to each other numerically as well. Not only that, but they actually appear to be within the ASCII range of characters, which would make sense since they are being used in a string comparison. Now, let's figure out what this value actually is! Sadly, when statically analyzing a binary we do not have the option of popping it open in a debugger and waiting for the value to be filled, so it will be best to just break apart the values by hand and determine what they are.

TIP

Most reverse engineering experts will frequently use pencil and paper when actually diving into a binary. Taking notes, quickly mapping paths or values, and many other things are always going to be faster on paper as long as we still have keyboards on computers. Try to always keep a pencil and paper next to you so you can quickly note addresses, values and other random items; our memories are never going to be perfect.

However, before we get further into it, here is a small anomaly within IDA's interpretation of this chunk of code that is interesting. In the middle of all our *mov* instructions, there is one *mov* that IDA has determined is writing to our input buffer.

```
.text:004011E4mov     [ebp+Dst], 74h
```

How could this be? Peeking back at the variables IDA has determined this function has, we can see it has attempted to automatically assign a different variable reference for every single *mov* instruction except for this anomaly. At a glance, this really does make the binary seem more complex and confusing. Because of the way IDA Pro disassembles functions, it sometimes has difficulty internally determining the actual structure of memory. Furthermore, it is a generally good rule of thumb not to totally rely on IDA Pro's dissemination of variables and arguments; instead, use them as references at a glance when required. This anomaly is due to these types of issues that IDA has; it hasn't appropriately taken into account stack alignment adjustments in its dissemination of the variables. So, ebp+dst, as IDA has determined it to be, would also be var_80 if IDA named it as such. We know a pointer was pushed into the stack prior to the function being called, and because of this IDA has not appropriately accounted for the stack shift inherent in this operation, thus rendering this misleading piece of code. Knowing how the stack is structured, we can determine how it looks as follows:

[Buffer Created]
[Pointer to Our Input String]

[Stack Pointer Save and Return Address]

[Real Input String]

Therefore, this instruction, which IDA says is overwriting our pointer, is actually writing the first byte of our character array, and thus is the first character of our password string. Now that we have taken note of that small problem and can appropriately account for it, we can sit down and figure out what this character array actually contains and what the password is. See Table 4.1.

NOTE

When performing static analysis on binaries, it never hurts to map out the stack in a table on paper as you work through a function. Especially in more complex functions, we can't completely rely on IDA to get it right every time. This is why understanding actual execution and stack structure is even more important in static analysis. We have to infer how the executable is going to behave every step of the way without having the ability to test and verify our assumptions.

Table 4.1 Password Dissection

Position	Value	Character
70	0	(NULL)
71	64	d
72	72	r
73	6F	o
74	77	w
75	73	s
76	73	s
77	61	a
78	70	p
79	79	y
7A	6D	m
7B	73	s
7C	69	i
7D	73	s
7E	69	i
7F	68	h
80	74	t

We can see now that this is in fact a valid ASCII string that is being compared against our input buffer. Since this is a generic ×86 binary, of course it is being assigned backwards in code. Therefore, our password is "thisismypassword". Going back and looking at the code, we can see that the conditional within this function is using the result of *strcmp()* in order to determine whether we provided the correct password or not. We have now passed the first hurdle in determining what this application does and how it is protecting itself.

Moving forward in the code, we can see that there is actually a check for the correct password using a conditional jump, followed by a final call to an external function prior to returning. Even more interesting, it seems that the return value of the called function, *sub_40100*, is passed through as the return value for this function as well.

```
.text:00401255        add    esp, 8
.text:00401258        call   sub_401000
.text:0040125Dloc_40125D: ; CODE XREF: input_process+83
.text:0040125D; input_process+85 ❑j
.text:0040125D        mov    esp, ebp
.text:0040125F        pop    ebp
.text:00401260        retn
```

How is this determined? Just prior to the call to the function *sub_401000*, the stack is actually shifted by 8 bytes. More specifically, ESP is incremented. This moves the stack appropriately so that the actual stack location of both functions' return values will be the same, thus passing the value through.

Stepping back for a moment, we now know a few new bits of information about the binary. First, we now know we need a password in order to progress anywhere. Secondly, if we do enter the correct password, it will progress to a second function, which is the final return value given to the entrypoint main function. In order to get a successful response, then, we will be required to meet whatever conditions exist within the third function as well. On that note, let's rename the call to *SecondCheck* and double-click it to view it.

```
.text:00401000 SecondCheck  proc near ; CODE XREF: input_process+98
.text:00401000 Dst   = byte ptr -4D0h
.text:00401000 var_450      = byte ptr -450h
.text:00401000 Dest  = byte ptr -400h
.text:00401000        push   ebp
.text:00401001        mov    ebp, esp
.text:00401003        sub    esp, 4D0h
.text:00401009        push   esi
.text:0040100A        push   edi
.text:0040100B        mov    ecx, 13h
.text:00401010        mov    esi, offset aPleaseSelectAn ; "Please select an option
from the follow"…
.text:00401015        lea    edi, [ebp+var_450]
```

```
.text:0040101B        rep movsd
.text:0040101D        movsw
.text:0040101F        movsb
.text:00401020        push   80h     ; Size
.text:00401025        push   0       ; Val
.text:00401027        lea    eax, [ebp+Dst]
.text:0040102D        push   eax     ; Dst
.text:0040102E        call   _memset
.text:00401033        add    esp, 0Ch
.text:00401036        push   80h     ; Size
.text:0040103B        push   0       ; Val
.text:0040103D        lea    ecx, [ebp+Dest]
.text:00401043        push   ecx     ; Dst
.text:00401044        call   _memset
.text:00401049        add    esp, 0Ch
.text:0040104C loc_40104C: ; CODE XREF: SecondCheck+1AB
.text:0040104C        mov    edx, 1
.text:00401051        test   edx, edx
.text:00401053        jz     loc_4011B0
.text:00401059        lea    eax, [ebp+var_450]
.text:0040105F        push   eax
.text:00401060        push   offset aS      ; "%s"
.text:00401065        call   printf
.text:0040106A        add    esp, 8
.text:0040106D        lea    ecx, [ebp+Dst]
.text:00401073        push   ecx
.text:00401074        push   offset a127s_0 ; "%127s"
.text:00401079        call   _scanf
.text:0040107E        add    esp, 8
.text:00401081        push   80h     ; Count
.text:00401086        lea    edx, [ebp+Dst]
.text:0040108C        push   edx     ; Source
.text:0040108D        lea    eax, [ebp+Dest]
.text:00401093        push   eax     ; Dest
.text:00401094        call   _strncat
.text:00401099        add    esp, 0Ch
.text:0040109C        push   offset Str2    ; "Exit"
.text:004010A1        lea    ecx, [ebp+Dst]
.text:004010A7        push   ecx     ; Str1
.text:004010A8        call   _strcmp
```

```
.text:004010AD        add    esp, 8
.text:004010B0        test   eax, eax
.text:004010B2        jnz    short loc_4010CD
.text:004010B4        lea    edx, [ebp+Dst]
.text:004010BA        push   edx
.text:004010BB        push   offset aOperationSComp ; "Operation: %s: Completed\n"
.text:004010C0        call   printf
.text:004010C5        add    esp, 8
.text:004010C8        jmp    loc_401195
.text:004010CD loc_4010CD:   ; CODE XREF: SecondCheck+B2
.text:004010CD        push   offset aSelect ; "Select"
.text:004010D2        lea    eax, [ebp+Dst]
.text:004010D8        push   eax     ; Str1
.text:004010D9        call   _strcmp
.text:004010DE        add    esp, 8
.text:004010E1        test   eax, eax
.text:004010E3        jnz    short loc_4010FE
.text:004010E5        lea    ecx, [ebp+Dst]
.text:004010EB        push   ecx
.text:004010EC        push   offset aOperationSCo_0 ; "Operation: %s: Completed\n"
.text:004010F1        call   printf
.text:004010F6        add    esp, 8
.text:004010F9        jmp    loc_401195
.text:004010FE loc_4010FE: ; CODE XREF: SecondCheck+E3
.text:004010FE        push   offset aDrop  ; "Drop"
.text:00401103        lea    edx, [ebp+Dst]
.text:00401109        push   edx     ; Str1
.text:0040110A        call   _strcmp
.text:0040110F        add    esp, 8
.text:00401112        test   eax, eax
.text:00401114        jnz    short loc_40112C
.text:00401116        lea    eax, [ebp+Dst]
.text:0040111C        push   eax
.text:0040111D        push   offset aOperationSCo_1 ; "Operation: %s: Completed\n"
.text:00401122        call   printf
.text:00401127        add    esp, 8
.text:0040112A        jmp    short loc_401195
.text:0040112C loc_40112C: ; CODE XREF: SecondCheck+114
.text:0040112C        push   offset aCreate ; "Create"
.text:00401131        lea    ecx, [ebp+Dst]
```

```
.text:00401137         push   ecx    ; Str1
.text:00401138         call   _strcmp
.text:0040113D         add    esp, 8
.text:00401140         test   eax, eax
.text:00401142         jnz    short loc_40115A
.text:00401144         lea    edx, [ebp+Dst]
.text:0040114A         push   edx
.text:0040114B         push   offset aOperationSCo_2 ; "Operation: %s: Completed\n"
.text:00401150         call   printf
.text:00401155         add    esp, 8
.text:00401158         jmp    short loc_401195
.text:0040115A loc_40115A: ; CODE XREF: SecondCheck+142
.text:0040115A         push   offset aExit_0       ; "Exit"
.text:0040115F         lea    eax, [ebp+Dst]
.text:00401165         push   eax    ; Str1
.text:00401166         call   _strcmp
.text:0040116B         add    esp, 8
.text:0040116E         test   eax, eax
.text:00401170         jnz    short loc_401188
.text:00401172         lea    ecx, [ebp+Dst]
.text:00401178         push   ecx
.text:00401179         push   offset aOperationSCo_3 ; "Operation: %s: Completed\n"
.text:0040117E         call   printf
.text:00401183         add    esp, 8
.text:00401186         jmp    short loc_401195
.text:00401188 loc_401188: ; CODE XREF: SecondCheck+170
.text:00401188         push   offset aInvalidCommand ; "Invalid command failure.
Please try aga"…
.text:0040118D         call   printf
.text:00401192         add    esp, 4
.text:00401195 loc_401195: ; CODE XREF: SecondCheck+C8, SecondCheck+F9
.text:00401195         push   80h    ; Size
.text:0040119A         push   0      ; Val
.text:0040119C         lea    edx, [ebp+Dst]
.text:004011A2         push   edx    ; Dst
.text:004011A3         call   _memset
.text:004011A8         add    esp, 0Ch
.text:004011AB         jmp    loc_40104C
.text:004011B0 loc_4011B0: ; CODE XREF: SecondCheck+53
.text:004011B0         mov    al, 1
```

```
.text:004011B2          pop     edi
.text:004011B3          pop     esi
.text:004011B4          mov     esp, ebp
.text:004011B6          pop     ebp
.text:004011B7          retn
.text:004011B7 SecondCheck    endp
```

As you can see, this function is larger than all the others. Although our techniques so far are promising, we are moving into a more complicated function and, as such, need to shift our method of analysis a bit. When diving into a function a bit more complex, it is better to get an overall picture of the function calls made and different conditionals that may exist prior to actually getting down and dirty with its operations. On that note, let's begin by mapping out the series of function calls made within this function. Specifically, we want to go into the graph view and see what IDA Pro has determined the layout of conditionals to be for us, allowing us to dissect much more information about the call structure between these functions, as shown in Table 4.2.

Table 4.2 Functions

memset()	Fills both stack buffers within the function with nulls (0×00)
printf()	Outputs command request header text
scanf()	Receives and stores user input in the first buffer
strncat()	Copies the received data from the first buffer to the second
strcmp()	Compares the input provided by the user against a static command string
printf()	Outputs the appropriate command result

As you can see, the overall structure of the function isn't as complicated as it seems, except for the multiple conditional statements that exist within it. However, from a glance at the graph view, reading the output that *printf* is specified to give on different conditions, it becomes obvious that this method is some form of a command parsing engine; it takes commands with *scanf*, parses to check for them, and then outputs the appropriate results. Additionally, when looking at this in greater detail, we can see by the graph view that this is in fact an infinite loop as well. Although we would be able to see this if we analyzed the jump statements within this binary long enough, IDA Pro provides us with this information by showing an extra overall wrapping connection from the final step of the function to the beginning. It becomes obvious that this is, in fact, a simple parsing loop for command strings. (See Figure 4.2.)

Figure 4.2 Reading *printf* Output in Graph View

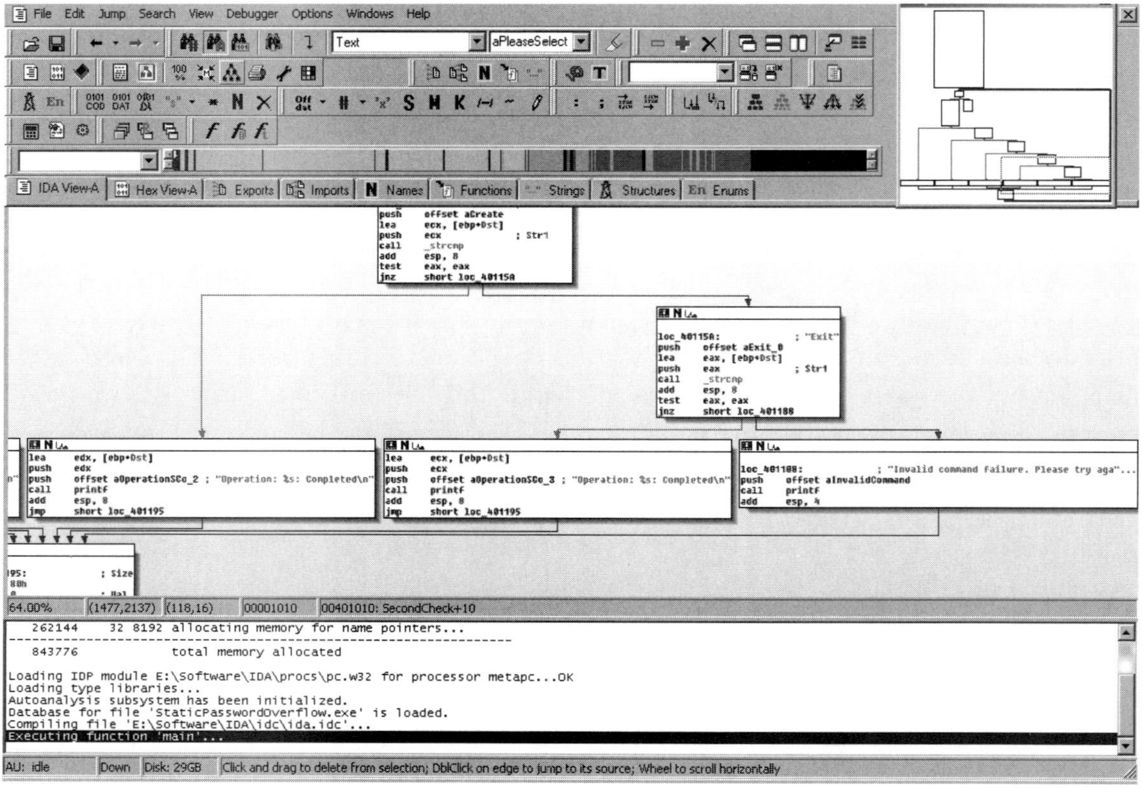

Solutions Fast Track

Understanding Execution Flow

- ☑ IDA has many tools and views, such as the graph view, which assist in rapidly assessing the actual operations within a binary.
- ☑ Always review the overall flow of a function or set of functions prior to diving into complete dissections of the entire block of code.
- ☑ It will help in the long run to obtain a level of comfort with other number sets; specifically with base-16 hex. The faster you are able to determine a value in your head, the less time you will spend with calculator open.
- ☑ Compilers will do funny things to binaries during compile time; these can sometimes be convoluted or pointless, but the differences are generally minimal in most optimization cases.

Recovering Hard Coded Password

- ☑ Identifying major conditional statements, such as final true and false results of comparisons is useful in many more settings than just hard coded passwords. However, beware of optimizers and obfuscation techniques which abuse jumps and comparisons in order to make the executable hard to read.
- ☑ Finding the root chunk of code which controls a true/false statement is a good critical step in identifying how to recover a password or bypass a major check within a binary.
- ☑ Developers and compilers alike do this in fickle manners sometimes for the sack of performance. Always be vigilant with what appears to be malignant portions of code; you never know what they might be for.

Frequently Asked Questions

Q: Why does IDA Pro have difficulty with identifying function arguments?

A: Statically analyzing code without context is a difficult task to perform. With typecasting and optimizing compilers, this problem is compounded. It will always be safer to check the callers of routines rather than rely on IDA Pro alone in identifying what exactly the arguments to a function are. Even so, a vigilant eye needs to be kept open for common compiler code which manipulates structures because in assembly they are just groups of values.

Chapter 5

Debugging

Solutions in this chapter:

- Debugging Basics
- Debugging in IDA Pro
- Use of Debugging while Reverse Engineering
- Heap and Stack Access and Modification
- Other Debuggers

☑ Summary

Introduction

Debugging is the act of locating bugs in software. Generally this is done by developers as bugs are worked out of their software. Debugging can take many forms. Beginning programmers often use output as a rudimentary form of debugging. This output can be *printf* statements in the case of C.

The most popular story on the use of debugging involves actual bugs. The origin of the use of "bug" to refer to a programming mistake is attributed to Admiral Grace Murray Hopper. A moth got caught in one of the relays from the Harvard University's Mark II computer. The removal of the moth was coined *debugging*.

Debuggers are programs themselves that run and monitor the execution of other programs. The debugger can control and alter the execution of the target program. Memory and variables can be monitored and altered as well.

Debugging Basics

Debuggers are an essential tool in the reverse engineer's toolbox. The ability to perform runtime analysis speeds up program understanding and reverse engineering. Certain tasks are easier within a debugger. Call chains can be watched instead of guessed.

Tracking indirect calls is much easier during debugging. A call through a register is an example of an indirect call. IDA Pro's static analysis tracks indirect calls in a very limited fashion. Cross references are not created.

Debugging allows us to watch, observe, and guide our reverse engineering. We do not want to reverse engineer entire programs, but rather the interesting parts.

Tools & Traps…

User mode vs. kernel mode debuggers

User mode debuggers operate on processes. They themselves are standard processes and as such are limited to what memory can be accessed. User mode debuggers cannot access kernel memory and thus are not useful to debug code operating in kernel mode. Kernel mode code can be the operating system, modules, and drivers.

Attacks on driver vulnerabilities are becoming more popular. An attack at the kernel level bypasses many modern protections. An example is the Broadcom Wireless Driver Probe Response SSID Overflow (CVE-2006-5882). A specially crafted probe allows arbitrary code execution in the kernel. The attack is at a level lower than the firewall. The packets never get processed by the firewall.

Continued

Rootkits have been developed as drivers for years. Some malware has adopted the rootkit strategies, embedding themselves in the kernel. DRM software often has drivers that contain vulnerabilities as evidenced by CVE-2007-5587, which was discovered being exploited in the wild.

Examples of user mode debuggers are IDA Pro and Ollydbg.

Under the Windows environment the de facto kernel mode debugger was Compuware's SoftICE included in DriverStudio. SoftICE has reached end of life and is no longer supported. Fortunately Microsoft has been making great strides with its debugging tools.

Breakpoints

Breakpoints stop execution of a program within the debugger at a location of our choosing. Execution is stopped and control is passed to the debugger. Breakpoints come in two different forms: hardware and software. Hardware breakpoints, as their name indicates, require specialized hardware support from the CPU.

Hardware Breakpoints

The IA-32 family of processors provides support for four hardware breakpoints. The hardware breakpoints use special debug registers. These registers contain the breakpoint addresses as well as control information and breakpoint type.

Breakpoint addresses are stored in debug registers D0 to D3. In order to set breakpoints a size field is needed. The possible sizes are 1, 2, or 4 bytes. Breaks on execution use a size of 1 byte. The possible sizes have been expanded to include 8 bytes for 64-bit CPUs. There are various conditions to trigger the breakpoints.

- Break on execution
- Break on memory access (reads and writes)
- Break on memory write only
- Break on I/O port access (rarely used, most debuggers do not have this as an option)

Software Breakpoints

Software breakpoints can only break on execution. A software breakpoint is simulated because of the lack of hardware support. A software breakpoint replaces the original instruction with an instruction that traps the debugger. In IA-32 processors, the new instruction is generally INT 3 (0xCC). The debugger must keep track of the original instruction.

When a software breakpoint executes, the INT 3 instruction passes control to the debugger. The debugger looks up the breakpoint in an internal table and replaces the INT 3 with the original instruction. The debugger then sets the instruction pointer back, making the saved instruction the next instruction to execute. The entire process is not visible to the user; the debugger will display disassembly with the original instructions in place.

Using Breakpoints

Software breakpoints are used by debuggers more than hardware breakpoints. The main reason is that there is not a set limit to software breakpoints. Some anti-debug techniques involve calculating checksums on code sections to determine if any instructions are changed. Anti-debug techniques are covered in detail in Chapter 6.

Hardware breakpoints can be set on memory, unlike software breakpoints. Breaking on memory access can allow us to look for use of tables or memory corruption.

Single Stepping

Single stepping is the process of executing a single instruction and then returning control to the debugger. The IA32 family of processors supports single stepping directly in the hardware. By executing a single instruction at a time, we have the ability to carefully monitor certain sections of code. However, it is impractical to debug an entire program using this method. Generally single stepping is used to understand select portions of code.

From the CPU perspective, the debugger sets the TF (Trap Flag) on the EFLAGS register. Upon the execution of an instruction a debug exception will be generated. This debug exception is caught by the debugger as an interrupt, INT 0x01.

NOTE

Most debuggers provide step commands. Generally they are called *step into* and *step over*. From a user point of view the only differences happen on certain instructions, namely *call* and *rep*.

When a call is encountered, a step into command will follow the call, while a step over will break on the instruction following the call. This is generally done by setting a breakpoint, not by single stepping till the return.

Watches

We need to keep track of variables. In source level debugging, variables are abstract named locations with values. Within assembly, variables are usually memory locations. The compiler can sometimes optimize a variable into a register.

Watches are a way to display variables or useful expressions. They are updated whenever control is passed to a debugger, such as a breakpoint or single stepping. A watch can be a simple variable such as *loop_counter* or an expression like *packet[offset *4]*.

Exceptions

Exceptions are used by programmers to catch errors. The following pseudo demonstrates an exception:

```
_try
{
        open(file)
}
_except
{
        printerror
}
```

The debugger can either stop on an exception or pass it on to the application. An exception does not necessarily mean something went wrong. Many times programs use custom exceptions. Custom exceptions are also a common anti-debugging technique.

In Windows the exception 0xc000000005 is an Access Violation. This means that process attempted to access an address that is not mapped. You may have seen advisories that show:

```
:
Exception C0000005 (ACCESS_VIOLATION reading [41414141])
```

The address 0x41414141 is not mapped to the process and is most likely part of an overwrite using As. We want the debugger to stop on access violations.

Tracing

Tracing is the process of executing a program and recording information along the way. The UNIX command *strace* runs an executable while intercepting all system calls including passed arguments.

```
user@redbull:~$ strace ls
execve("/bin/ls", ["ls"], [/* 31 vars */]) = 0
brk(0) = 0x805c000
access("/etc/ld.so.nohwcap", F_OK) = -1 ENOENT (No such file or directory)
mmap2(NULL, 8192, PROT_READ|PROT_WRITE, MAP_PRIVATE|MAP_ANONYMOUS, -1, 0) =
0xb7eec000
access("/etc/ld.so.preload", R_OK) = -1 ENOENT (No such file or directory)
open("/etc/ld.so.cache", O_RDONLY) = 3
```

Tracing can be performed within a debugger, recording various levels of detail. Instruction level tracing is the most detailed. After executing each instruction, register values are recorded. This level of detail is not needed and the process is very slow.

Function tracing can set breakpoints at the entry of all functions or single step till a call. When a function is called, the execution is stopped. The debugger will record the arguments to the function and optionally can record data such as registers and the caller. Execution then resumes.

While function tracing is much faster than instruction tracing, it sometimes does not provide the necessary detail. Ideally we want to trace basic blocks. Basic blocks are sequential instructions that are executed without taking a branch. This type of tracing helps determine why certain branches are taken and, if needed, how to modify input in order to take different branches. While basic block tracing is much faster than instruction tracing, it is still slow. Starting with the P6 line of processors, Intel included hardware support to trace branches. Newer processors have more functionality in this area, but they all use MSR registers.

Intel has documented Last Branch Recording in Chapter 18 of the *Intel® 64 and IA-32 Architectures System Programming Guide*. New research in this area has been published and proof of concept tools have been released. See www openrce. org/blog/view/535/Branch_Tracing_with_Intel_MSR_Registers.

Debugging in IDA Pro

IDA Pro comes with a built-in debugger, which was introduced in version 4.50. The debugger is implemented as a plug-in. This is a true testament to the extensibility of IDA Pro.

IDA isn't limited to debugging the local system. The debugger can operate locally as well as remotely over the network. The debugging clients allow IDA to debug other machines running different operating systems and even CPUs. Authentication is available, but best practices suggest debugging only over the local network.

IDA Pro supports the following debugging environments:

- Win32 Local
- Win32 Remote
- Win64 Remote
- Linux Remote (x86 only)
- OSX Remote (x86 only)
- WinCE Remote (ARM only)

NOTE

IDA Pro can change register values from the GUI. However, memory locations cannot be changed from the GUI.

IDC, IDA Pro's scripting language, must be used to change memory locations. Memory locations include data and executable code. The IDC functions such as *PatchByte()*, *PatchWord()*, and *PatchDword()* must be used.

The debugger menu option is made available if the binary being analyzed matches one of the previous targets listed. Debugger settings are available from the debugger menu option as shown in Figure 5.1.

Figure 5.1 Debugger Setup Options

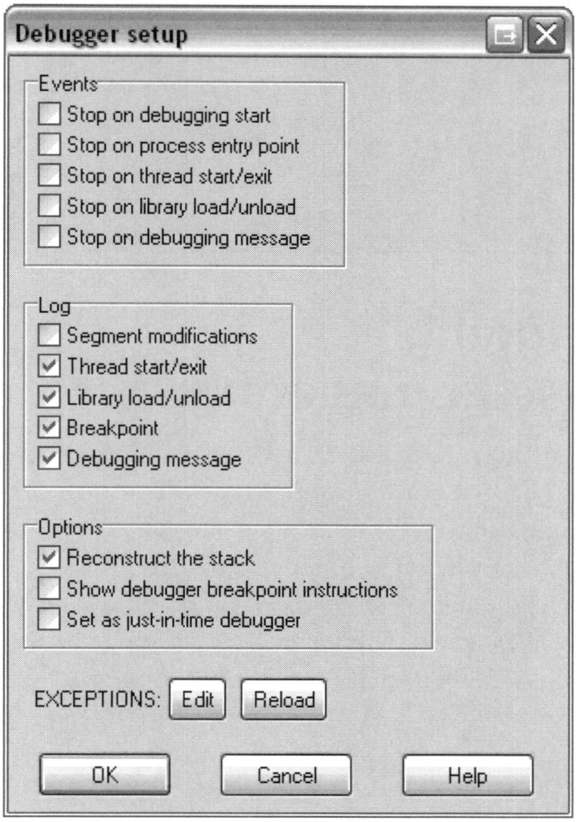

Some of the notable options are:

- Stop on debugging start – This option will stop before the entry point in the case of a PE binary with a TLS section.
- Stop on entry point – This option stops at the listed entry point. Some initialization may have been completed.
- Set as just-in-time debugger – Windows allows a debugger to be set as default when programs crash.
- Exceptions – This option controls how IDA handles exceptions, whether they are passed to the application or not.

Tools & Traps...

Writing your own IDA Pro debugger client

The IDA SDK allows you to write plug-ins. Included in the SDK is the source code to the Linux Debugger plug-in/client. The plug-in's source code shows the interface to work with IDA.

Using the provided source code a plug-in/client could be written for an operating system or architecture not currently supported.

Use of Debugging while Reverse Engineering

In order to demonstrate debugging within IDA Pro we will use Netcat as an example. Netcat is a network tool, whose name comes from the combination of network and the UNIX command *cat*. You can pipe data to and from other programs over a network, which is why it is known as the "TCP/IP Swiss army knife."

In December 2004, vulnerability was reported in Netcat for Windows 1.1 (www.vulnwatch.org/netcat/netcat-111.txt). We will be analyzing the vulnerable version 1.1 (http://packetstormsecurity.org/UNIX/netcat/nc11nt.zip), while the patched version is 1.11 (www.vulnwatch.org/netcat/).

NOTE

Netcat is a networking tool capable of allowing remote access. Some antivirus vendors classify it as a hacking tool. Please be aware of this and take the necessary precautions.

The vendor advisory describes a remote buffer overflow when using the "-e" option. The vulnerability resides in the *SessionWriteShellThreadFn* function within the dosexec.c source file. Select parts of the vulnerable function are shown in the following code snippet:

```
static VOID
SessionWriteShellThreadFn(LPVOID Parameter)
```

```
{
      PSESSION_DATA Session = Parameter;
      BYTE RecvBuffer[1];
      BYTE Buffer[BUFFER_SIZE];
      BYTE EchoBuffer[5];
      DWORD BytesWritten;
      DWORD BufferCnt, EchoCnt;
      DWORD TossCnt = 0;
      BOOL PrevWasFF = FALSE;
      BufferCnt = 0;

      //Loop, reading one byte at a time from the socket.
      while (recv(Session->ClientSocket, RecvBuffer, sizeof(RecvBuffer), 0) != 0)
           {
              EchoCnt = 0;

              Buffer[BufferCnt++] = EchoBuffer[EchoCnt++] = RecvBuffer[0];
              if (RecvBuffer[0] == '\r')
                   Buffer[BufferCnt++] = EchoBuffer[EchoCnt++] = '\n';

              //Trap exit as it causes problems
              if (strnicmp(Buffer, "exit\r\n", 6) == 0)
                   ExitThread(0);
              //
              //If we got a CR, it's time to send what we've buffered up down to the
              //shell process.
              if (RecvBuffer[0] == '\n' || RecvBuffer[0] == '\r') {
                  if (! WriteFile(Session->WritePipeHandle, Buffer, BufferCnt,
                  &BytesWritten, NULL))
                  {
                   break;
                  }
                  BufferCnt = 0;
              }
      }
      ExitThread(0);
}
```

The function contains a receive loop with two possible exits. The first exit occurs if the buffer contains an *exit\r\n* string. The second exit requires either a *\n* or *\r* as the received byte and a failure for the WriteFile call.

NOTE

Netcat sends a *n* as a newline when run from Windows. The command to terminate is *exit\r\n*. In order to terminate the newline needs to be sent as *r**n*.

After opening nc.exe in IDA Pro, the debugger tab is available. If an executable format is supported by one of the debuggers, the debugger tab will be visible.

Process options need to be configured. Specifically, we need to configure the command line arguments. The vulnerability is only present when the *-e* option is used. In order to use this option we must also supply the *-l* (listening) options as well as the *-p* (port number option). The *-e* option executes a program passing any input it receives over the network. We don't need the executed program to do anything, so we can use the *more* program. Figure 5.2 shows typical process options using our command line arguments.

Figure 5.2 Debugger Application Setup

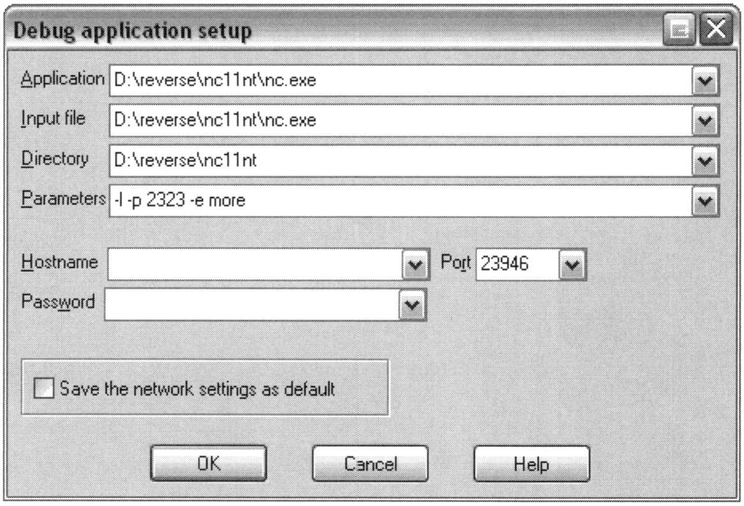

NOTE

If we are debugging a dll, the setup is slightly different. The dll would be in the input file box, while the application that uses the dll would go in the application box.

This type of setup is very common with Internet Explorer. The iexplore.exe binary does very little work and leaves all the heavy lifting to dlls.

Debugger hotkeys are:

- F9 Start debugger/Continue process (if already debugging)
- F2 Set/Remove breakpoint
- F7 Step into
- F8 Step over
- CRTL + F7 Run until return
- F4 Run to cursor
- CTRL + F2 Terminate process

We find the *SessionWriteShellThreadFn* function by looking at the imports for WriteFile. From the cross references we determine that the function address is .text:00401520. We then set a breakpoint at the beginning of the function. Figure 5.3 shows the graph view of the function. The function has been renamed to *SessionWriteShellThreadFn* and stack variable *buf* has been renamed to *RecvBuffer* for readability.

Figure 5.3 SessionWriteShellThreadFn Function

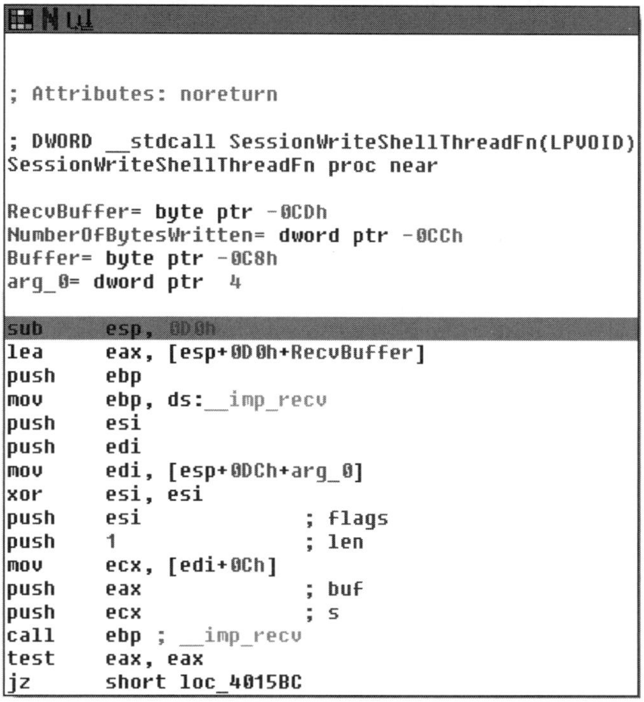

```
; Attributes: noreturn

; DWORD __stdcall SessionWriteShellThreadFn(LPVOID)
SessionWriteShellThreadFn proc near

RecvBuffer= byte ptr -0CDh
NumberOfBytesWritten= dword ptr -0CCh
Buffer= byte ptr -0C8h
arg_0= dword ptr  4

sub     esp, 0D0h
lea     eax, [esp+0D0h+RecvBuffer]
push    ebp
mov     ebp, ds:__imp_recv
push    esi
push    edi
mov     edi, [esp+0DCh+arg_0]
xor     esi, esi
push    esi             ; flags
push    1               ; len
mov     ecx, [edi+0Ch]
push    eax             ; buf
push    ecx             ; s
call    ebp ; __imp_recv
test    eax, eax
jz      short loc_4015BC
```

Push F9 and the debugger will start. The different windows will rearrange themselves. The debugger is running nc.exe using the passed arguments. Since there hasn't been a connection yet, our breakpoint hasn't been hit.

Start a cmd.exe shell and we will use another instance of Netcat to connect to the debugged one. This Netcat will be called the Netcat client in order to differentiate between the debugged Netcat. Use the command line:

```
nc localhost 2323
```

At this point we will hit our breakpoint. The register window will contain values similar to Figure 5.4. A stack window will be displayed similar to Figure 5.5.

Figure 5.4 Registers

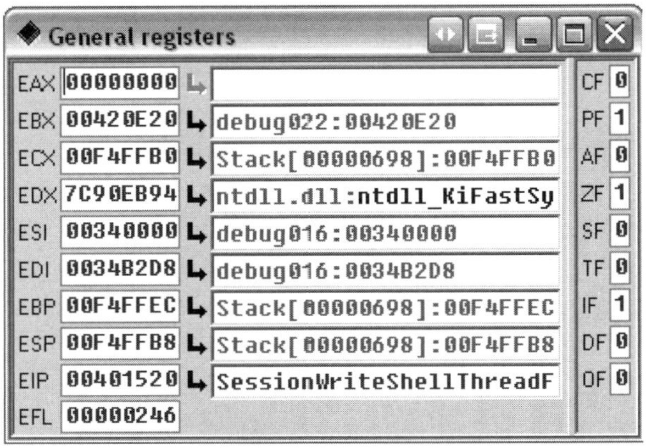

Figure 5.5 Stack

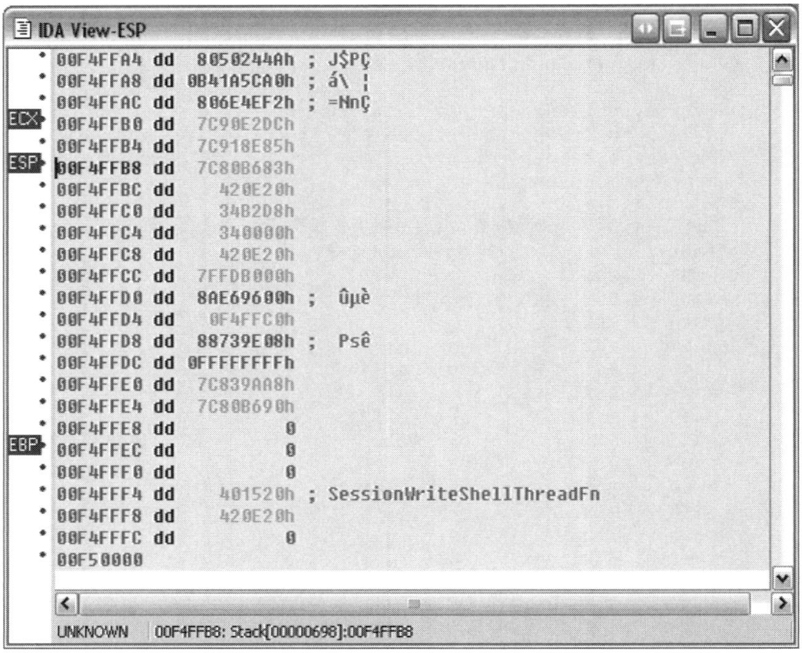

The register window shows the register values on the left and the right side contains any interpretation of the values. Registers can be changed by either right-clicking in the value box or typing directly into the value box. New views are available by right-clicking on registers and most addresses. The views can be assembly or hex.

Stepping with the F8 key (step over) will avoid going into the actual *recv* call. The *recv* call will block until it receives data. Type **hello\n** into the Netcat client. The *recv* call will return now that it has received data. We can continue single stepping.

1. The basic block in Figure 5.6 does the following:
2. Reads the single byte from the *recv* call into register *al*
3. Writes the byte into the buffer
4. Increments a buffer counter in register *esi*

Figure 5.6 Reading and Writing Bytes to the Buffer

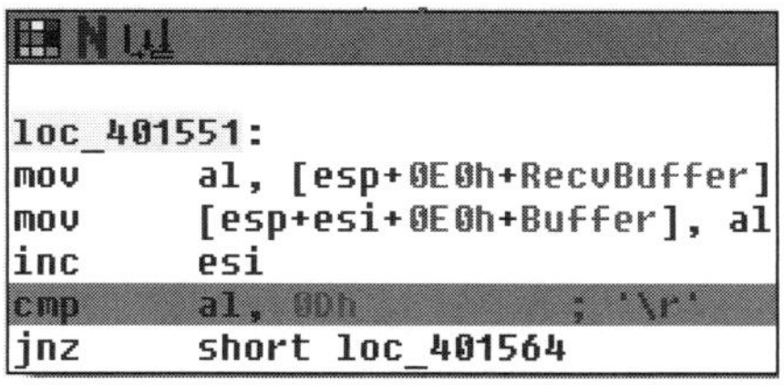

We can set another breakpoint on the instruction *cmp al, 0x0d* in the basic block from Figure 5.6. Pressing F9 (continue) runs the program until another breakpoint is hit. Execution will stop at the breakpoint we just set. Notice the value of *esi* has incremented after each character. The character can still be seen in register *al*. Right-click the breakpoint and select **disable breakpoint**.

After the entire command from the Netcat client has been copied into the buffer, WriteFile is called. Figure 5.7 shows the basic block containing the call. Set a breakpoint on *test eax, eax*, which is the instruction following the WriteFile call. When we continue (F9), the debugger will stop on this instruction.

Figure 5.7 Basic Block Containing the Call

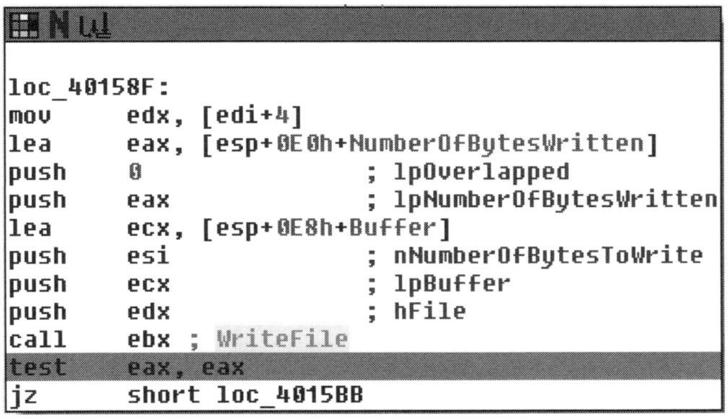

We know the size of the buffer; it is 0xc8 bytes according to Figure 5.7. However, the stack looks different than expected since this function is called by CreateThread().

We can set a conditional breakpoint on the instruction *cmp al, 0x0d* from the basic block shown in Figure 5.8. In order to set a conditional breakpoint, right-click on the disabled breakpoint and select **Edit breakpoint**. Enter **esi == 0xc8 || esi == 0xcc** in the condition box as in Figure 5.8. The breakpoint will hit at the last location of the buffer and then upon overwriting the next DWORD.

Figure 5.8 Conditional Breakpoint

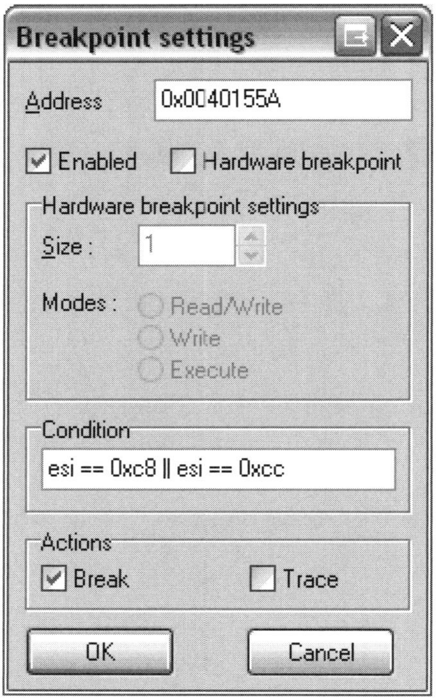

WARNING

The condition box takes an IDC statement, which is evaluated. A common mistake is using a single equal = (assignment), when a double == (evaluation) is needed. This is a classic bug type in C code.

We need to send more data. From the stack view, it appears that 272 bytes will write past the end of the page. Other machines or operating systems may have different memory layout. The easiest way to send the data is to build a string in a text editor and then paste it into the Netcat client.

Figure 5.9 Stack

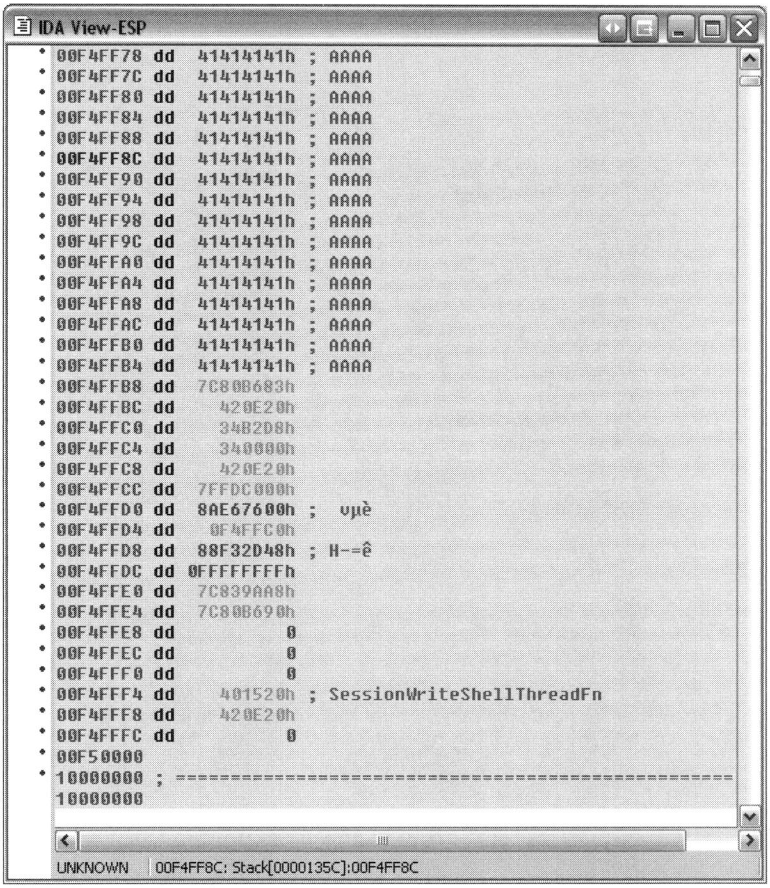

After the breakpoint hits, the stack looks like Figure 5.9. There is still data to be read and there is very little space left on the page. The next breakpoint is the first of the stack corruption. When the program is allowed to continue, we see the warning and then the exception will come up as shown in Figure 5.10.

Figure 5.10 Exceptions

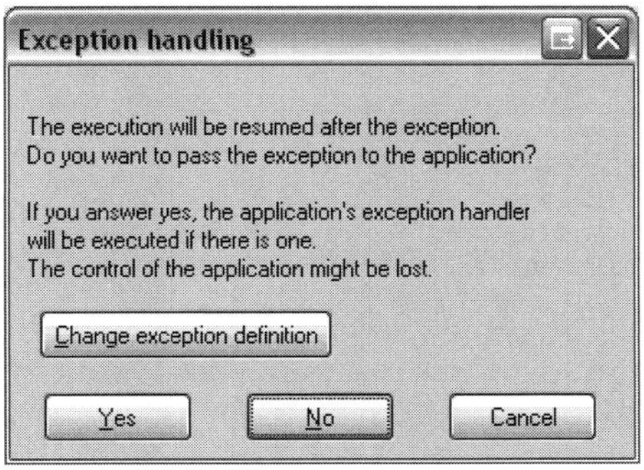

The exception is configured to stop the program. We can go into **Change exception definition** and select **Pass to application**. See Figure 5.11.

Figure 5.11 Handling Options

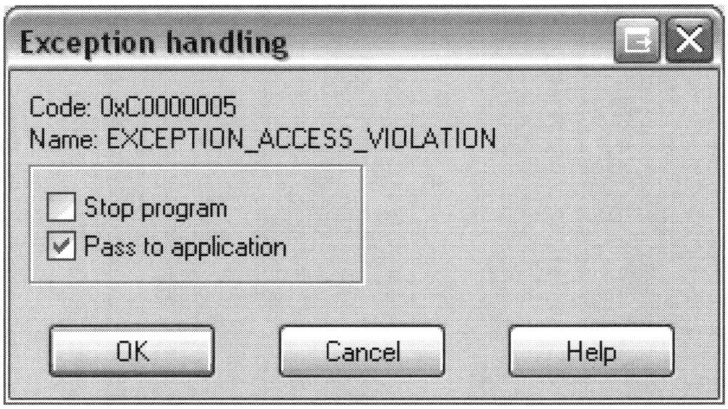

The program will terminate with this exception. IDA Pro has recorded the exception in the log window:

nc.exe: The instruction at 0x401555 referenced memory at 0xF50000. The memory could not be written (0x00401555 -> 00F50000)

```
Debugger: Thread terminated: id=00001660 (exit code = 0xC0000005).
Debugger: Thread terminated: id=000017AC (exit code = 0xC0000005).
Debugger: Process terminated (exit code = C0000005h).
```

Heap and Stack Access and Modification

Detecting memory corruption is important to the reverse engineer. Stack and heap overflows are attacks that overwrite and corrupt memory.

The debugger can be used to detect memory corruption. Some ways of detecting corruption can be done manually while others are more applicable to being scripts or plug-ins.

Microsoft began adding stack cookies to their compiler beginning with Visual Studio 2003, using the GS command line switch. At the entry of a function, a stack cookie is placed on the stack. The cookie is calculated by taking a global security cookie, __security_cookie, and XORing it with the *esp* register. During an exit of the function, the stack cookie is XORed with the *esp* register. The result of the operation should be *__security_cookie*. This value is passed to the *__security_check_cookie()* function. If the passed value matches *__security_cookie*, then *__security_check_cookie()* returns allowing the original function to continue as designed. A more detailed explanation is available here: http://uninformed.org/index.cgi?v=7&a=2&p=1.

The idea of the protection is that the cookie check will fail if the stack has been corrupted. If the check fails, the process will be terminated with an exit code of 0xc0000409. In order to catch the stack corruption, we can set a breakpoint in *__security_check_cookie()*, as shown in Figure 5.12. Alternately, the breakpoint can be set directly on the *__report_gsfailure()* function. The *__security_check_cookie()* function is compiled in statically and the address will change depending on the binary.

Figure 5.12 Setting a Breakpoint

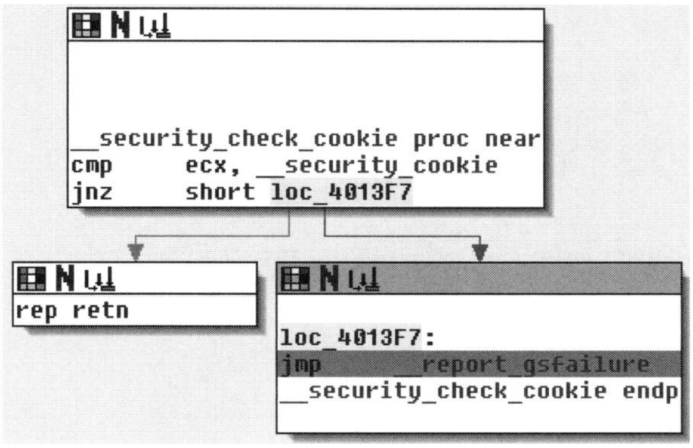

Checking for heap corruption is more dependent on the operating system being used. Windows XP SP2, Windows 2003, and Vista have various methods of heap protection built in. However, unlike the GS stack protection, the heap protections are part of the operating system. Various techniques have been developed to pass heap protections, but data from a fuzzer will most likely be caught by these protections.

There are multiple checks and simple breakpoints may not be sufficient. Debugger based scripting or a plug-in would be ideal. The heap functions can be hooked to provide allocation data. The protection functions can be hooked to report corruption. For systems without such thorough protection functions, hooked functions containing checks could be added. Rather than stopping attacks, the checks notify us of corruption as soon as possible.

WARNING

Debuggers can change the environment and behavior of a process.

Processes started from a debugger use a debug heap, unlike starting the process normally. Attaching to a process is not affected. This difference is important when looking for heap corruption. In order to disable the use of the debug heap, set the environment variable *_NO_DEBUG_HEAP* to 1.

```
set _NO_DEBUG_HEAP=1
```

Microsoft's gflags.exe utility allows the setting of many debugging options. gflags.exe is part of Debugging Tools for Windows (www.microsoft.com/whdc/devtools/debugging/default.mspx).

Operating system code can have debugger checks. kernel32.Unhandled Exception Filter alters its behavior based on the presence of a debugger. This behavior was originally mentioned in Dave Aitel's paper "MSRPC Heap Overflow – Part II" and subsequently in the *Shellcoder's Handbook*.

Other Debuggers

The debuggers within IDA Pro are very useful. You have full access to the static analysis, renamed functions, and other parts of code that have been reverse engineered. Like any type of tool, people have preferences for different tools. There are different debuggers available to the reverse engineer.

Each debugger has advantages and disadvantages. It usually comes down to a matter of personal preference. The following paragraphs provide a brief overview of some other debuggers.

Windbg

Debugging Tools for Windows is a collection of debuggers from Microsoft (www.microsoft.com/whdc/devtools/debugging/default.mspx). There are two different versions available from Microsoft, a 32-bit and 64-bit version.

Debugging Tools for Windows 32-bit Version runs on:

- Windows NT 4.0
- Windows 2000
- Windows XP (32-bit or 64-bit)
- Microsoft Windows Server 2003 (32-bit or 64-bit)
- Windows Vista (32-bit or 64-bit)
- Windows Server 2008 (32-bit or 64-bit)

Debugging Tools for Windows 64-bit Version runs on:

- Windows XP (64-bit)
- Microsoft Windows Server 2003 (64-bit)
- Windows Vista (64-bit)
- Windows Server 2008 (64-bit)

The 64-bit version of the Debugging Tools for Windows should only be used if debugging native 64-bit applications.

Windbg is the debugger you will most likely use, although other debuggers are included, such as NTSD, CDB, and KD. Windbg is a user and kernel mode debugger. One of its primary benefits is the tight integration with Windows.

Ollydbg

Ollydbg is a free win32 user mode debugger, available from www.ollydbg.be. Although source code isn't available, there is an SDK provided. Many plug-ins have been written for Ollydbg.

Ollydbg is a very popular debugger among reverse engineers. It was written from the reverse engineer standpoint. Some notable plug-ins include scripting, anti-anti-debugging, and tracing. There are many tutorials available for Ollydbg ranging from basic reversing to security bug hunting to breaking software protections.

The current version is 1.10 and is no longer supported as the author is working on the upcoming 2.0 release full time. There have been various vulnerabilities reported in the debugger, including a format string vulnerability (CVE-2004-0733). Packers use these vulnerabilities in order to prevent debugging. However, reversers have released plug-ins which patch these vulnerabilities, one of the most popular being Olly Advanced (www.openrce.org/downloads/details/241/Olly_Advanced).

Immunity Debugger (Immdbg)

Immdbg is a free debugger released by Immunity Inc. (www.immunityinc.com/products-immdbg.shtml). When you first run Immdbg, you will notice that it is ad supported. These are not ads for the next hot stock, but rather they are ads purchased by companies looking for security talent. If Immdbg looks similar to Ollydbg, it is not by mistake. Immunity Inc. licensed the source code to Ollydbg in order to add features useful for exploit development.

Having a source license allows them to fix bugs. New features include graphing, a command line, and remote debugging. The standout new feature is the built-in Python scripting. Some sample scripts are included to demonstrate the Python API. Other scripts have been released by users on the Immdbg forum.

PaiMei/PyDbg

PaiMei is a reverse engineering framework (http://paimei.openrce.org/). It is written in Python and has scripts to use analysis from IDA Pro. One of PaiMei's key components is PyDbg. PyDbg is a scriptable debugger written in Python allowing it to integrate with IDAPython (http://d-dome.net/idapython/).

IDAPython is an IDA Pro plug-in that allows scripting. It wraps many of the IDC and SDK functions. Unfortunately IDA Python does not wrap many of the debugger calls. However, IDAPython can use PyDbg in order to debug and combine runtime analysis with IDA Pro's static analysis.

GDB

All the other debuggers discussed thus far have been for Windows operating systems. The GNU Project Debugger (GDB) is available for most UNIX systems. GDB is primarily a source level debugger. However, GDB can also operate at the assembly level.

GDB uses a text-based interface, although numerous graphical front ends have been developed. They communicate with GDB using MI (Machine Interface). Scripting languages can drive GDB by using MI.

Tools & Traps…

Packers

What is a packer? Packers are most commonly used in Win32 environments. They compress executables and uncompress the image in memory when executed. Thus, in order to analyze the binary, the actual uncompressed image is needed. A common open source packer is UPX, http:///upx.sourceforge.net. UPX is designed to operate in both directions; it is able to restore a packed binary back to the original binary. Most packers are not designed this way and there are modifications often made to UPX to prevent the symmetric behavior.

Often the binary is run within a debugger or emulated environment until the original entry point (OEP) is determined. Memory is then dumped along with an appropriate PE header. Generally the imports are destroyed, so the imports are added back in. This is the basic method for extracting the original image. There are variants as the packing/unpacking arms race continues.

Packers are used in many other programs besides malware. The goal of the packer is to make reverse engineering more difficult, while also lowering the file footprint. Some of the anti-debugging/reversing techniques will be discussed in Chapter 6. Software protection makes use of packers. This can include software from shareware to commercial packages. Any software dealing with DRM will also typically use packers and/or anti-debugging. Skype, a popular telephony program, makes substantial use of these techniques. In order to evaluate or binary audit packed software for vulnerabilities, the unpacked image is needed.

Summary

IDA Pro's debugger is very powerful and allows for much greater program understanding than static analysis alone. The debugger can operate locally as well as remotely with the most common operating systems.

A benefit to using the IDA Pro debugger over other debuggers is the availability of any reverse engineering work we have done. This includes renaming functions, tables, and local variables.

There are times when IDA Pro's debugger is not the best solution. Various other debuggers are available.

Chapter 6

Anti-Reversing

Solutions in this chapter:

- Debugging
- Example Overview
- Obfuscation

☑ Summary

Introduction

Anti-debugging is a natural occurrence that should be expected; as soon as people started reversing applications it was only a matter of time before other people started trying to make it harder, implausible or even impossible for someone to reverse their application. Anti-debugging, like reverse engineering or coding in assembly, is an art form. The trick of course is to try to stop the person reversing the application. However, in most instances these attempts range from the absurdly lame to the truly difficult. At first it may be your presumption that only malicious software would seek to impede your reversing progress, but really you will find it everywhere and indeed there are legitimate jobs out there just for people to create such anti-reversing technologies, especially in the video game industry. In this chapter, what we hope to do is write a fairly comprehensive overview of anti-debugging and anti-disassembly techniques. Make no mistake—these tricks and techniques are designed to make your life and job more difficult, and in some instances there really is no good solution to getting around the problem presented. However, it is important to know and remember one thing: Given enough time and motivation, the reverse engineer always wins.

First, if we really want to understand anti-reversing, then a little knowledge of exactly how debugging and disassembling is done would be helpful. This of course is not meant to be an all-encompassing perspective on the art, but rather a brief introduction to how it works, with the intent of using that knowledge as leverage to understanding anti-reversing techniques.

Debugging

To really understand debugging, a brief tour of the various interrupts and debug registers is necessary, especially in regards to what state changes occur in the process. The *Intel Software Developers Manual* is once again the best place for reference in this regards, as it goes much deeper into details than I can. However, basically the IA-32 platform handles debugging through one of a couple of means.

First, the debug registers. There are eight debug registers supporting the ability to monitor up to four addresses. The registers themselves are accessed through variants of the MOV instruction with the debug registers potentially serving as either the destination or source operands. It should be known that accessing the registers is a privileged process requiring ring-0 privileges, which of course is a limiting feature but considering the power they give it makes sense. For each breakpoint, it is necessary to specify the address in question, the length of the location (ranging between a byte and a dword), a handler when a debug exception is generated and finally whether this breakpoint is even enabled. The first three debug registers, DR0 through DR3 can contain three 32-bit addresses that define the address where breakpoints should occur. The next two debug registers, DR4 and DR5 respectively, have alternating roles depending on mode of operation. When the debugging extensions (DE) flag is set in control register 4 (CR4),

DR4 and DR5 are reserved and cause an invalid-opcode exception when an attempt to reference them is made. If the flag is unset, then DR4 and DR5 are instead aliases for debug registers 6 and 7.

Debug register 6 (DR6), is also known as the debug status register and indicates the results of conditional checks at the time of the last debug exception. DR6 is accessed as a bit pattern, with bits zero through three being related to the first three debug registers. Each of these bits indicates which breakpoint condition was met and caused a debug exception to be generated. Bit 13 of DR6 when set indicates that the next instruction references a debug register and is used in conjunction with a portion of DR7, which we will describe momentarily. Bit 14 is perhaps the most interesting for our purposes; it indicates when set that the processor is in single-step mode, which is yet another concept we will introduce momentarily. Finally in use is bit 15, which indicates that a debug exception was raised as a result of a task switch when the debug trap flag was set. Finally, we arrive at debug register 7, or DR7 as you might have guessed. This is a very interesting register to hackers of all kinds as it's also known as the debug control register and like DR6 is interpreted as a bit field. The first byte of this register corresponds to whether a breakpoint is active, and if so its scope. Bits zero, two, four and six determine whether a debug register is enabled or not, with bits one, three, five and seven corresponding to the same breakpoints but on a global scope. The scope in this instance is defined as whether the breakpoint persists through task switches, with globally enabled breakpoints being available to all tasks. In later versions of the processor, according to the Intel manual, bits eight and nine are not supported. However traditionally they provide the ability to determine the exact instruction that caused the breakpoint event. Next we have bit 13; this is an interesting bit as it allows for breaking before accesses to the debug registers themselves. Finally we have bits 16 through 31. These bits determine what types of access cause a breakpoint, and what the length of the data at the address is. When the DE flag in CR4 is set, bits 16 to 17, 20 to 21, 24 to 25 and 28 to 29 are interpreted in the following manner:

```
00 - Break on execution
01 - Break on write
10 - Break on I/O read or writes
11 - Break on read and writes but not instruction fetches
```

However, when the DE flag is not set the interpretation remains the same except for values of 10 which are undefined. Bits 18 to 19, 22 to 23, 26 to 27 and 30 to 31 correspond to the lengths of the various breakpoints with a value of 00 indicating that the length is 1, and 01 indicating a 2-byte length. 10 is undefined on 32-bit platforms, with it indicating a length of 8 bytes on 64-bit processors. Finally, as you might have deduced, a value of 11 indicates that the length in question is 4 bytes in length.

Now it may seem a little confusing trying to tie all these bit sequences together with DR0 to DR3, but it really isn't. Each 2-bit combination corresponds to a given sequential register in the range of DR0 through DR3. These lengths must be aligned on certain boundaries dependent on their size—for instance 16-bit values need to be on word boundaries and 32-bit ones on double-word boundaries. This is enforced by the processor by masking the relevant low-order bits of the address; thus an unaligned address will not yield performance as expected. An exception is generated if any addresses in the range of the starting address plus its length are accessed, effectively allowing for unaligned breakpoints by using two breakpoints; each breakpoint is appropriately aligned and between the two of them they cover the length in question. Now one last note of interest here—when the breakpoint access type is execution only, then the length specified should be set to 00; any other value results in undefined behavior.

NOTE

Interestingly enough, the debug registers have not received tremendous amounts of attention publicly. However, privately there are numerous and quite effective rootkits and backdoors that make use of them. For instance, if so inclined, a person could hide a process in a linked list of processes by setting a global access breakpoint on the pointer to their process structure. When an access to that address occurs, a debug exception occurs and they can redirect into their handler and perform any number of tasks, including returning the address of the next process in the list.

To make matters worse, they can enable the GD flag in DR7 and cause accesses to the debug registers themselves to have an exception generated, thwarting even attempts to inspect the registers to check for their current configuration.

Now we've mentioned debug exceptions throughout the description of the registers, but we haven't really done anything beyond mention them. The IA-32 processor has an interrupt vector specified in the interrupt descriptor table which was described previously in Chapter 2. Of these, the processor dedicates two interrupt vectors to these exceptions. These two interrupts are one and three, which are the debug and breakpoint exceptions, respectively. The debug exception, or INT 1, is generated by multiple events and DR6 and DR7 should be consulted to determine what type of event occurred exactly. In the process of an exception there are two general classes, faults and traps. Essentially, the difference between the two classes is whether the instruction that generated the interrupt was executed or not by the time the handler gets control. In faults, control is handed to the handler first, whereas in traps execution control is passed to the handler after the instruction that caused the exception is generated. Of these two classes, we have several different conditions that fall into the classes: instruction breakpoint, data and I/O breakpoint, general-detect, single-step and task-switch conditions. Of these the instruction breakpoint, general-detect and arguably task-switch conditions are fault class, while data and I/O breakpoint and single-step conditions are trap class.

Instruction breakpoint class conditions are the highest priority exceptions and occur when an instruction at an address referenced in DR0 to DR3 is attempted to be executed. We say that these exceptions are the highest priority meaning that they receive service first; however, there are instances where these events may not even be triggered We'll delve into that a bit later though. Now if you recall in the description of the EFLAGs register earlier, there was the resume flag in present. The problem is that it's possible for a debug exception to be re-raised as a result of this exception being a trap-class exception. This is where the resume flag comes into play, as it prevents looping of the debug exception. We'll cover this also shortly when we start to delve into specific methodology. The other fault class condition is the general-detect condition; this is raised when the relevant bit in DR7 is set, protecting access to the debug registers.

Data and I/O breakpoints are trap-class conditions and are caused by data accesses of addresses in DR0 to DR3. Data accesses are essentially any condition that's not an execution attempt. The IA-32 processor contains an interesting quirk in that trap-class events occur after the instruction that caused them was executed. For instance, suppose you set a write breakpoint on address X whose value is 0 and then the application modifies address X, setting its value to 1. When the exception handler receives control, the value at address X will be 1, not 0. The Intel manuals suggest that applications that want to be able to interact with the original value should save it at the time of the breakpoint, although doing so creates an interesting but off-topic synchronization issue. Even more, these breakpoints, like instruction breakpoints, are not always exact; for instance, repeated execution of certain SIMD instructions can cause the exceptions to be raised until the end of the second iteration.

Single-step exceptions are also trap-class conditions, and are one of the more common conditions encountered when debugging. These conditions are caused when the trap flag in the EFLAGs register is set. Just like every other type of exception, single stepping comes with its own quirks. For instance, generally speaking the trap flag is not modified in the process of performing various tasks; however, certain instructions like software interrupts and INTO instructions do clear the trap flag. This effectively means that, in order to maintain control, a debugger has to emulate these instructions and cannot directly execute them, that is if they wish to continue inspiring single-step exceptions. Finally, task-switch exceptions occur after a task switch if the trap flag in the new tasks TSS is set; the exception is raised after the task-switch but prior to the first instruction in the new task.

Now, in addition to INT 1, there is also the breakpoint exception or INT 3. The breakpoint exception is interesting in that it allows for extension of breakpoints past the number supported by the debug registers. However, it requires modifying memory, an Achilles heel that we will talk about exploiting later. For now, all we really need to know is that it exists.

Upon hearing all of this, an inspired reader with a good imagination might already begin to see the conditions that could be checked for and ways that interruptions might be avoided. However, people don't purchase nonfiction books to inspire their imagination but rather to learn facts, so we'll cover some of the complications that can occur in the rest of the chapter. We will explore mostly ring-3 based implementations of circumventing both disassembly and debugging, but to spice things up some we will also touch on some ring-0 based concepts.

Example Overview

Because this is a somewhat tricky subject which can be difficult to explain exactly, especially within the constraints of a single chapter, we're going to take the following approach. I've written a simple application, a silly network RPC server and client. What we're going to do is take this (the server side component) application and harden it to reversing a bit, or rather do as much as we can within the constraints. Thus, we will have two products: the original program and the one we've hardened. The idea here is that we're going to reverse the anti-RCE process and, by making one ourselves and looking at the generated code, when you do eventually run across this sort of stuff, you'll have some concept of what's going on and how to get around it. The example codes we will be using for this chapter are available for download from the Syngress website. We will not be looking at the client side of the application, and it is there simply for you to experiment with, and for me to verify that everything is working as expected.

So without any further ado, let's take a brief look at the application prior to doing anything to it, so you have some idea of what it started off looking like. Figure 6.1 shows the relevant section of the control-flow graph that's generated when you go to **View > Graphs > Control Flow**, or if you hit **F12**, the options hotkey. This isn't the full graph, as it's not necessary and trying to fit it into a page was problematic. Basically, I want you just to get an idea of what we're looking at. However, I highly encourage you to download the source, compile it and look it over.

Figure 6.1 Control Flow Graph

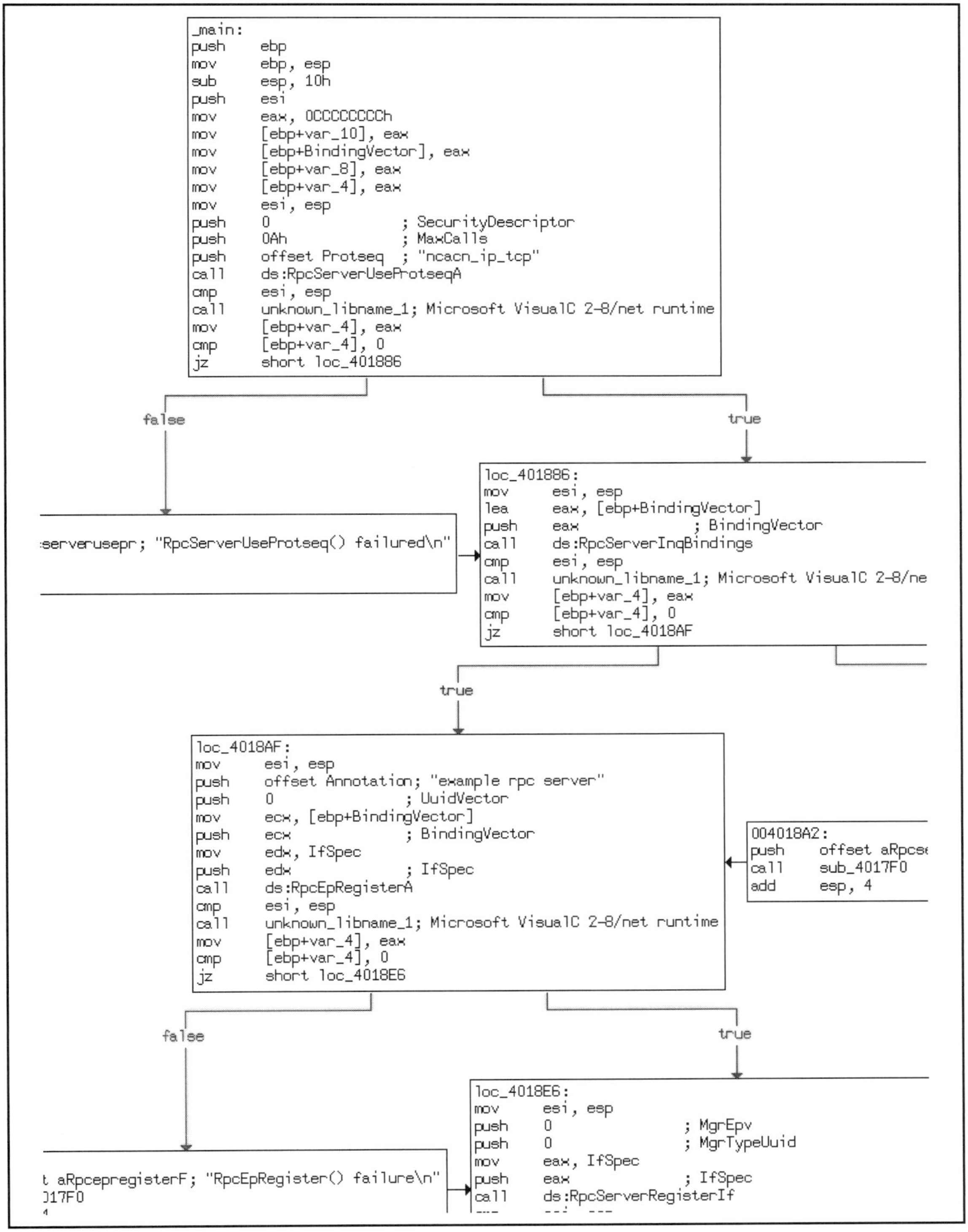

As you can see, this is obviously an application dealing with RPC in some form, and when we take a closer look it becomes clear that it's intended to be the server component, as one can tell just by the API calls. As you examine the code, take note that there are multiple string constants, that there is really no attempt to obfuscate what the application is doing and so on. Now, let's change that! If you feel a little lost at this point, pull the code down and take a closer look at it to get a better feel for it, as this is a chapter about anti-reversing, not about coding an RPC server and I really can't delve into the details much further.

Obfuscation

As we noted earlier, there is very little doubt about what is going on in this program, or at least what appears to be going on. (One should be careful about forming conclusions in regards to the application's purpose with such a small preview; however, in this case, what we see is what we get.) Let's start off with one of the simpler, although more effective, techniques and obfuscate the code a bit.

These types of techniques are common and really are something you should just accept as the normal routine; sans any other types of security, they generally won't pose much of a problem for you. What we're actually trying to do here is raise the bar as to who can read the code. You see, a decent percentage of people calling themselves "reverse engineers" or working as "incident responders," really aren't. Some of them can do little more than extract readable strings from the binary, others just read API calls, while some will take a guess of intent based on data in the imports section. An often-used trick—especially when dealing with packed binaries—is to modify some aspect of the binary so that if someone tries to dump the image straight from memory it will be corrupted.

Taking this all into account, let's take a look at the newly modified main routine in Figure 6.2.

Figure 6.2 Modified Main Routine

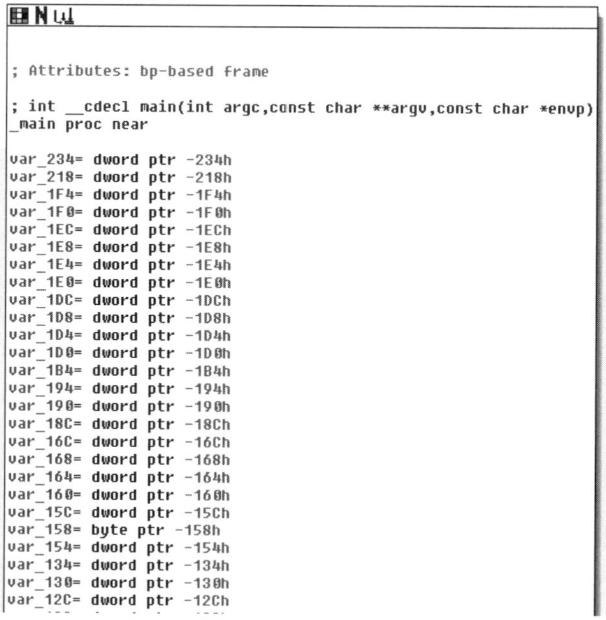

```
; Attributes: bp-based frame

; int __cdecl main(int argc,const char **argv,const char *envp)
_main proc near

var_234= dword ptr -234h
var_218= dword ptr -218h
var_1F4= dword ptr -1F4h
var_1F0= dword ptr -1F0h
var_1EC= dword ptr -1ECh
var_1E8= dword ptr -1E8h
var_1E4= dword ptr -1E4h
var_1E0= dword ptr -1E0h
var_1DC= dword ptr -1DCh
var_1D8= dword ptr -1D8h
var_1D4= dword ptr -1D4h
var_1D0= dword ptr -1D0h
var_1B4= dword ptr -1B4h
var_194= dword ptr -194h
var_190= dword ptr -190h
var_18C= dword ptr -18Ch
var_16C= dword ptr -16Ch
var_168= dword ptr -168h
var_164= dword ptr -164h
var_160= dword ptr -160h
var_15C= dword ptr -15Ch
var_158= byte ptr -158h
var_154= dword ptr -154h
var_134= dword ptr -134h
var_130= dword ptr -130h
var_12C= dword ptr -12Ch
```

Looking at just the entry point reveals that this new binary could be significantly more complex, just by the number of local variables now present. Of course, we are also looking at different views of the code, so this isn't quite as obvious as when we reach the first sets of instructions (see Figure 6.3).

Figure 6.3 First Instructions

```
var_C= dword ptr -0Ch
var_8= dword ptr -8
var_4= dword ptr -4
argc= dword ptr  8
argv= dword ptr  0Ch
envp= dword ptr  10h
arg_34= dword ptr  3Ch
arg_70= dword ptr  78h

push    ebp
mov     ebp, esp
sub     esp, 348h
mov     eax, dword_404000
xor     eax, ebp
mov     [ebp+var_C], eax
push    ebx
push    esi
push    edi
mov     eax, ds:dword_403118
mov     [ebp+var_6C], eax
mov     ecx, ds:dword_40311C
mov     [ebp+var_68], ecx
mov     dx, ds:word_403120
mov     [ebp+var_64], dx
mov     al, ds:byte_403122
mov     [ebp+var_62], al
mov     ecx, ds:dword_403124
mov     [ebp+var_C4], ecx
mov     edx, ds:dword_403128
mov     [ebp+var_C0], edx
mov     eax, ds:dword_40312C
mov     [ebp+var_BC], eax
mov     ecx, ds:dword_403130
mov     [ebp+var_B8], ecx
mov     edx, ds:dword_403134
mov     [ebp+var_B4], edx
mov     al, ds:byte_403138
mov     [ebp+var_B0], al
mov     ecx, ds:dword_40313C
mov     [ebp+var_A4], ecx
mov     edx, ds:dword_403140
mov     [ebp+var_A0], edx
```

The trend in change of the code continues as we note that, where once there was the start of RPC calls, we now have a long series of byte copies from the data segment onto the stack. You may note that the stack is declining while the data index is increasing; this is typical of things like string constants initializing a stack variable. We'll come back and look at this in a moment. However, before we delve into the changes I want you to be able to see and get a feel for the differences in the program, as seen in Figure 6.4.

Figure 6.4 Differences in the Program

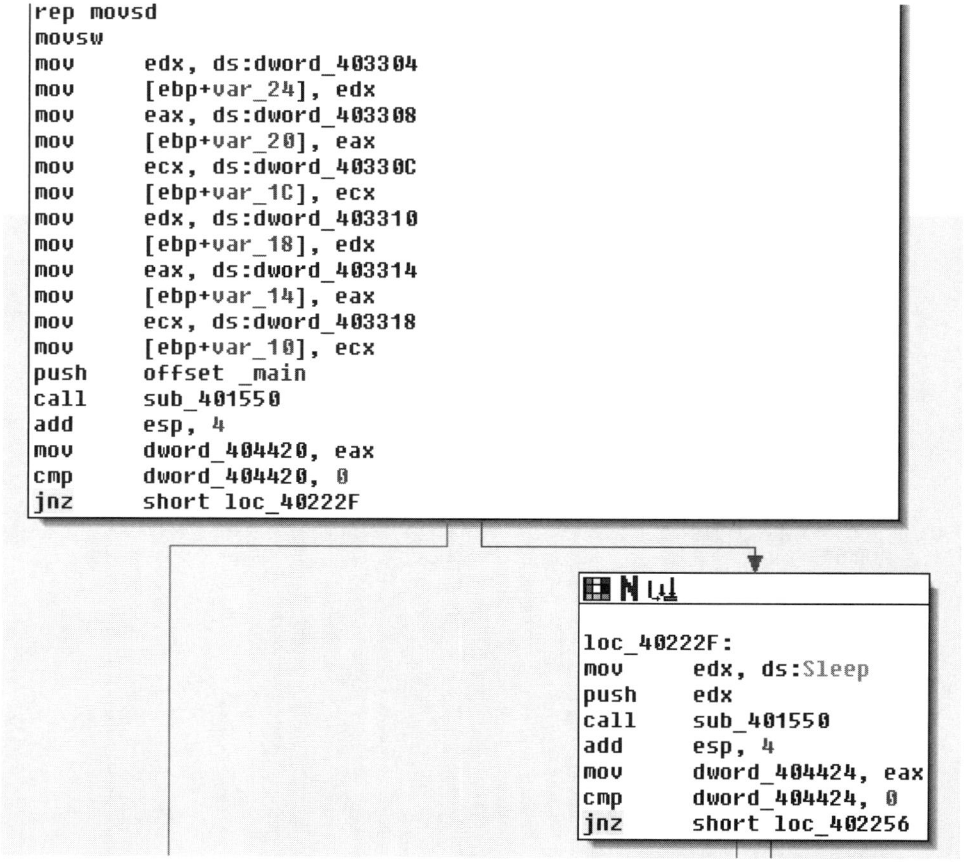

Continuing down the code, we finally get to the first "real" portion of code, which is a call to the function *sub_40155*. There are two calls to it, the first passing the address of the main function in as an argument, the second passing in the address of the kernel32 function *Sleep()*. Now that we have some idea of what we're getting ourselves into, let's take a bird's-eye view of the code (Figure 6.5) and see how crazy the structure of the application might be.

Figure 6.5 Complicated Structure of Application

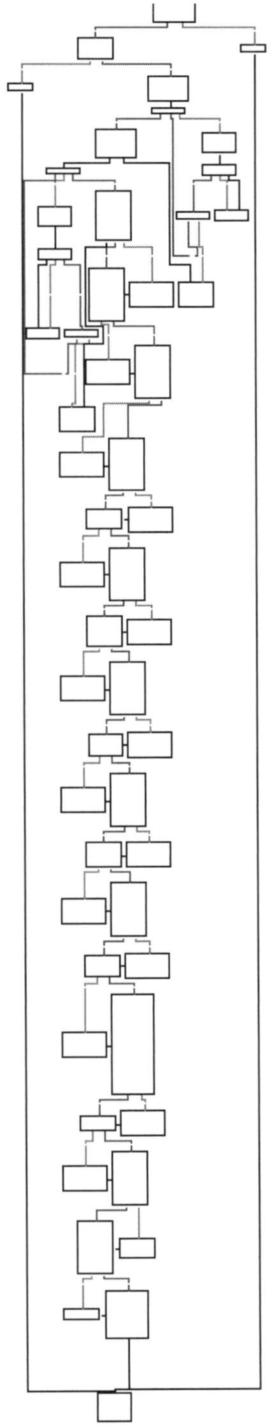

And here we have it—no matter how awkward the code initially appears, it exhibits fairly typical structures, still showing a basic Boolean logic to its overall composition. Most of the complexity appears to be towards the start, and then we see a simple series of if () { if () […] else […] } else [..] structures. Notice the first error branches to the right, whereas the second error branches off to the left, continuing on down to process termination.

Great! Now that we have some idea of what we're working with, let's go back and get a better idea of how exactly it changed and see if we can figure out how this little puzzle fits together, starting with that long series of copies from the data section to the stack. Starting at the beginning of the data, we have address loc_403118, so let's jump there and see what we can determine (Figure 6.6).

Figure 6.6 address loc_403118

```
.rdata:00403118 dword_403118    dd 2D23190Eh            ; DATA XREF: _main+16↑r
.rdata:0040311C dword_40311C    dd 3B021856h            ; DATA XREF: _main+1E↑r
.rdata:00403120 word_403120     dw 233Ch                ; DATA XREF: _main+27↑r
.rdata:00403122 byte_403122     db 0                    ; DATA XREF: _main+32↑r
.rdata:00403123                 align 4
.rdata:00403124 dword_403124    dd 0C23192Eh            ; DATA XREF: _main+3A↑r
.rdata:00403128 dword_403128    dd 3A5A5E47h            ; DATA XREF: _main+46↑r
.rdata:0040312C dword_40312C    dd 3A171A22h            ; DATA XREF: _main+52↑r
.rdata:00403130 dword_403130    dd 2B260F7Eh            ; DATA XREF: _main+5D↑r
.rdata:00403134 dword_403134    dd 1E561F43h            ; DATA XREF: _main+69↑r
.rdata:00403138 byte_403138     db 0                    ; DATA XREF: _main+75↑r
.rdata:00403139                 align 4
.rdata:0040313C dword_40313C    dd 0C23192Eh            ; DATA XREF: _main+80↑r
.rdata:00403140 dword_403140    dd 3A5A5E47h            ; DATA XREF: _main+8C↑r
.rdata:00403144 dword_403144    dd 2E0A0622h            ; DATA XREF: _main+98↑r
.rdata:00403148 dword_403148    dd 3B27146Ch            ; DATA XREF: _main+A3↑r
.rdata:0040314C dword_40314C    dd 2C401459h            ; DATA XREF: _main+AF↑r
.rdata:00403150 byte_403150     db 0                    ; DATA XREF: _main+BB↑r
.rdata:00403151                 align 4
.rdata:00403154 dword_403154    dd 1A23192Eh            ; DATA XREF: _main+C6↑r
.rdata:00403158 dword_403158    dd 38497E52h            ; DATA XREF: _main+D2↑r
.rdata:0040315C dword_40315C    dd 3A103C39h            ; DATA XREF: _main+DE↑r
.rdata:00403160 word_403160     dw 3C5Ch                ; DATA XREF: _main+E9↑r
.rdata:00403162 byte_403162     db 0                    ; DATA XREF: _main+F7↑r
.rdata:00403163                 align 4
.rdata:00403164 dword_403164    dd 0C23192Eh            ; DATA XREF: _main+103↑r
.rdata:00403168 dword_403168    dd 3A5A5E47h            ; DATA XREF: _main+10E↑r
.rdata:0040316C dword_40316C    dd 38011D22h            ; DATA XREF: _main+11A↑r
.rdata:00403170 dword_403170    dd 3A3D0E47h            ; DATA XREF: _main+126↑r
.rdata:00403174 dword_403174    dd 413342h              ; DATA XREF: _main+131↑r
```

As we examine the data that initially is referenced in the *main()* function, it becomes obvious that they are not ASCII characters. However, they appear to be NULL terminated, further supporting the idea of string constant initialization of local variables. Furthermore, we can take note that the data has some repeating patterns, for instance, look at how many sequences start with the bytes 0x0C2319. This may be indicative of weak encryption. We really won't be able to tell for sure until we get to a section of code that interacts with this data. We'll shelve the idea for now, but we can speculate that the program once accomplished a great deal with RPC and had many string constants, and now contains a bunch of random NULL terminated data that appears to at least have a repeating first three characters in most of the cases.

Now, knowing that there is not a lot we can do at the present time about the apparently encrypted data, unless we're willing to track down where the data gets used, we'll move on to the next portion and let the order of operations occur sequentially. Next up, we have the function *sub_401550* (Figure 6.7).

Figure 6.7 Function *sub_401550*

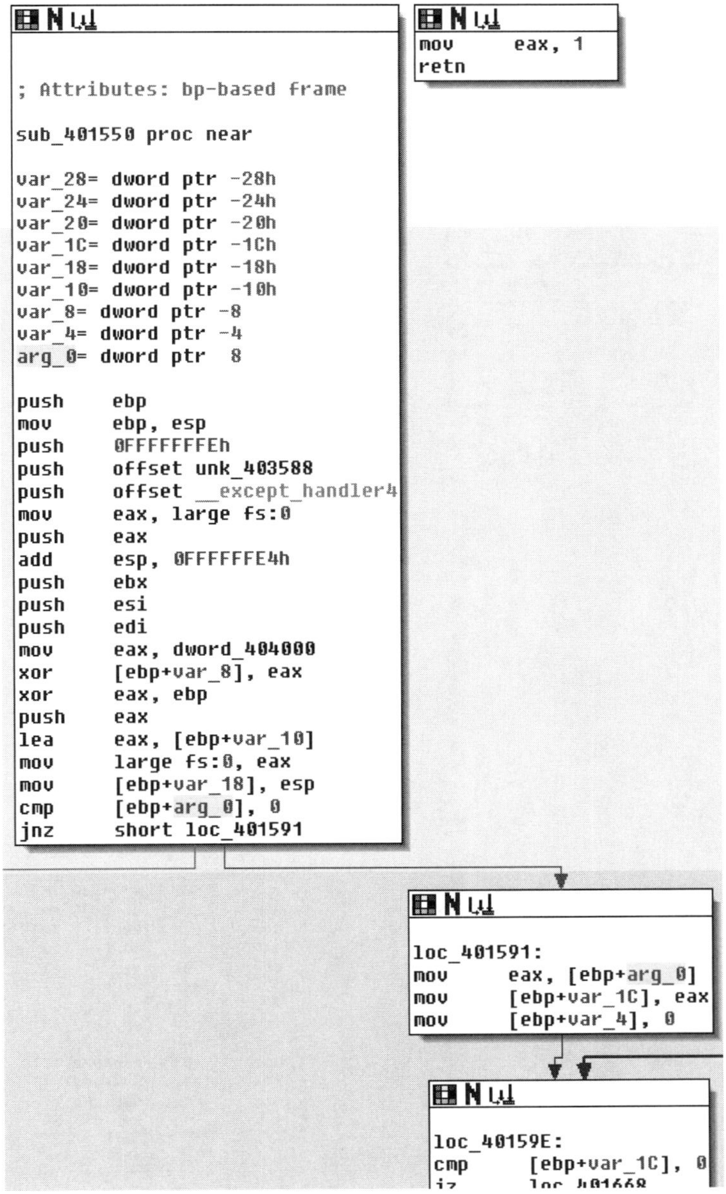

As we examine the entrance of *sub_401550*, we see a function that's a bit more normal looking than *main()*. As you may recall, in the first part of *main()* this is invoked twice, once with an argument of the address of main and once with a pointer to *Sleep()*. The user portion of the code starts at the cmp instruction testing arg_0 against 0 towards the end of the first box. One thing to note in the compiler-generated code is that there is an exception handler setup.

So the first real thing we see is that the argument is tested against 0, or rather the pointer is checked against NULL, if it is, we branch off to the left presumably towards an abnormal termination. If the pointer is non-NULL then in the branch to the right, at loc_401591 we see the variable *var_1C* is initialized to the value of the routines parameter and *var_4* is initialized to 0 (Figure 6.8).

Figure 6.8 *var_4* Is Initialized to 0

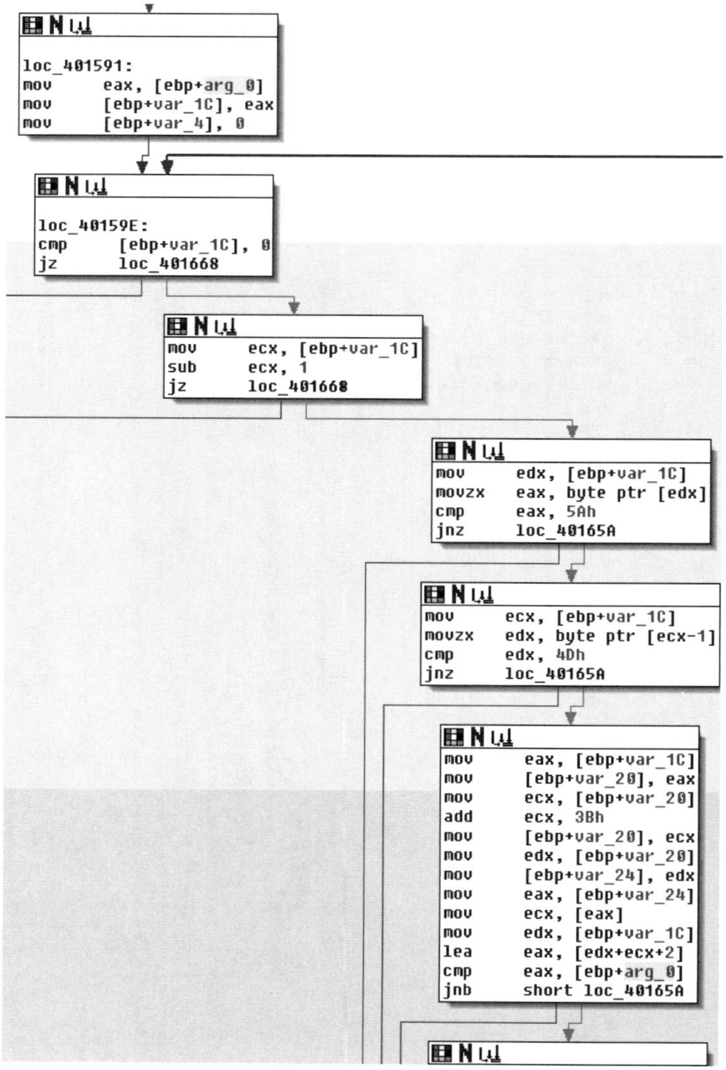

Continuing down through the function, we arrive at *loc_40159E*, which appears to also start the mark of a loop. This loop is made more obvious if you're looking at the function in its entirety in the graphs; just remember this as being the loop back point, as you can tell by the arrow coming from the upper right back to *loc_40159E*. As far as functionality, what we see in this box is that the pointer copy of the routine parameter is tested against NULL again and then if it is, a branch occurs to the left, otherwise the routine branches to the right. From there we move down to the third box where the *var_1C* pointer is decremented by one and tested against zero, again branching to the left if the condition is true. From there, in the fourth box down we see the pointer dereferenced and checked against the byte 0x5A; if a match is found, then it moves to the fifth box, if not it branches to the left.

If there was a match, the byte prior to this is checked against the value 0x4D, meaning that we're looking for the 16-bit sequence 0x4D5A. As you may recall, this corresponds to the ASCII letters MZ, which is how a DOS header starts. Without looking further, we can make an educated guess that the function appears to be walking backwards through memory attempting to find the beginning of the file—this also explains the use of SHE as in theory it would be possible to touch bad memory and crash. We'll continue looking through the routine though; it would be bad practice to just presume based on so little.

Next, in the final visible box, if a match against 0x4D5A was found, the pointer is copied into *var_20*, and then incremented by 0x3B and this new pointer is copied into *var_24*. Finally, what we see next is the *var_1C*, or the pointer to the MZ summed with the value from *var_24* (*var_1C* + 0x3B) plus two. This incidentally corresponds to the offset of the PE header in the DOS header and it would appear that the final box attempts to find the start of the PE header and then compares that pointer to *arg_0* and branches dependent on result, presumably towards another abnormal exit (see Figure 6.9).

Figure 6.9 Match Against 0x4D5A

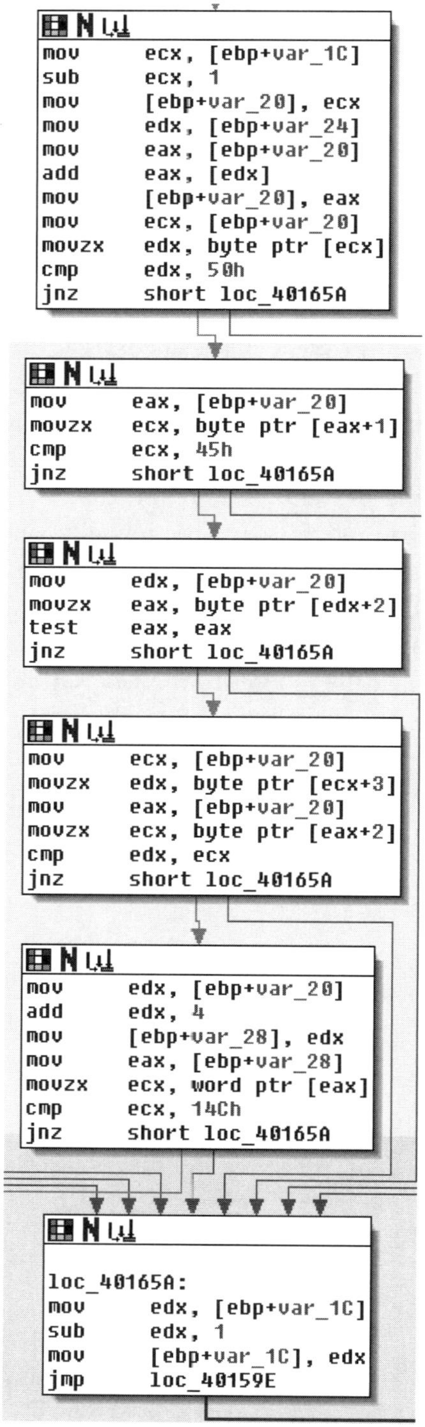

Moving into the next section, we see the same theme continued; in the first box we see a comparison against 0x50 followed by a check for 0x45. This in turn means it's looking for 0x4550. If there is a match, the next two boxes check to see if the sequence is followed by two zero bytes, meaning a full match thus far requires the sequence 0x45500000, or PE\0\0, which is of course the magic value for a PE header. Finally, as the last check in this section we see that they test against the value 0x14C, which corresponds to IMAGE_FILE_MACHINE_I386. All of this and previous sections make sense and it appears that the argument to the routine is a pointer that is an offset into an executable image. Given that input, the routine walks backwards through memory attempting to identify the start of the DOS header. Once found, it uses this information to locate the PE header and performs other light verifications of the data.

As we can see in the second to last section of code, if a total match is found, we branch off to the left, and if not then we jump to the location loc_40165A, which takes the pointer, decrements it by one and repeats the loop. Now that we know the body of the routine, let's examine the branches we haven't yet looked at and also take a look at the return value.

Figure 6.10 Jump to loc_401668

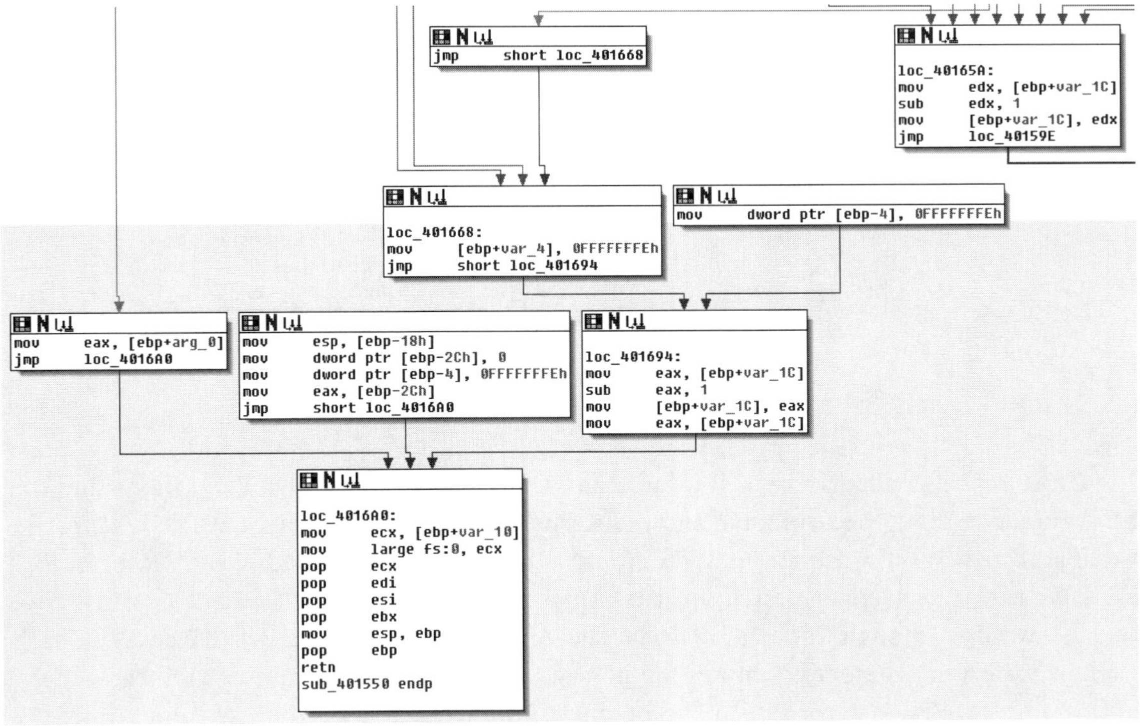

In Figure 6.10, starting from the top right, at loc_40165A, we see that if a match was indeed found then we jump to loc_401668, which modifies *var_4* and then passes control to loc_401694 which retrieves the base address, decrements it by one and then hands control to the cleanup portions of the routine. This means that the return value is a pointer to the base address of the image in question. We also see, on the far left, the result of the initial test of arg_0, which if it was NULL is the value returned. Thus we can conclude that, given a pointer, this routine finds its base address, returning that pointer or NULL on error.

Now, after reviewing this routine, let's take a look at the *main()* routine again and see what becomes apparent. We've renamed this routine to *FindBaseAddr()* as that is the functionality it provides (Figure 6.11).

Figure 6.11 *FindBaseAddr()*

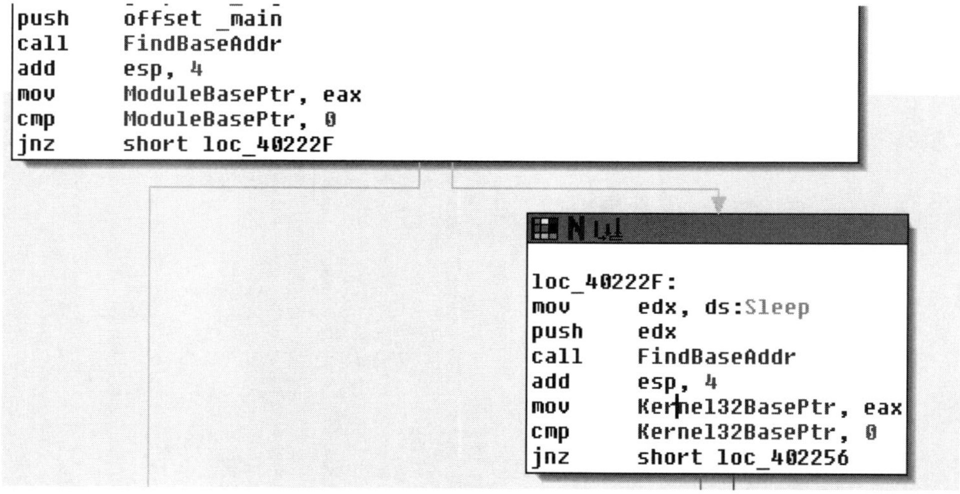

Updating this information in IDA, we can plainly see that these two invocations find the base address of the current module and the Kernel32 module, respectively. We've updated the variable names to reflect this, calling the pointers the *ModuleBasePtr* and *Kernel32BasePtr* for each. We now can make a guess at what is going on, as it's not typical for code to manually find its own base through this method, and totally unnecessary to find Kernel32. We can probably safely bet this is an attempt at obfuscating library calls. Following the routine's return, we see both pointers tested against NULL, branching to the left if this is true. Let's not consider these error branches yet, and examine the scenario where both calls to *FindBaseAddr()* succeed. (See Figure 6.12).

Figure 6.12 loc_402256

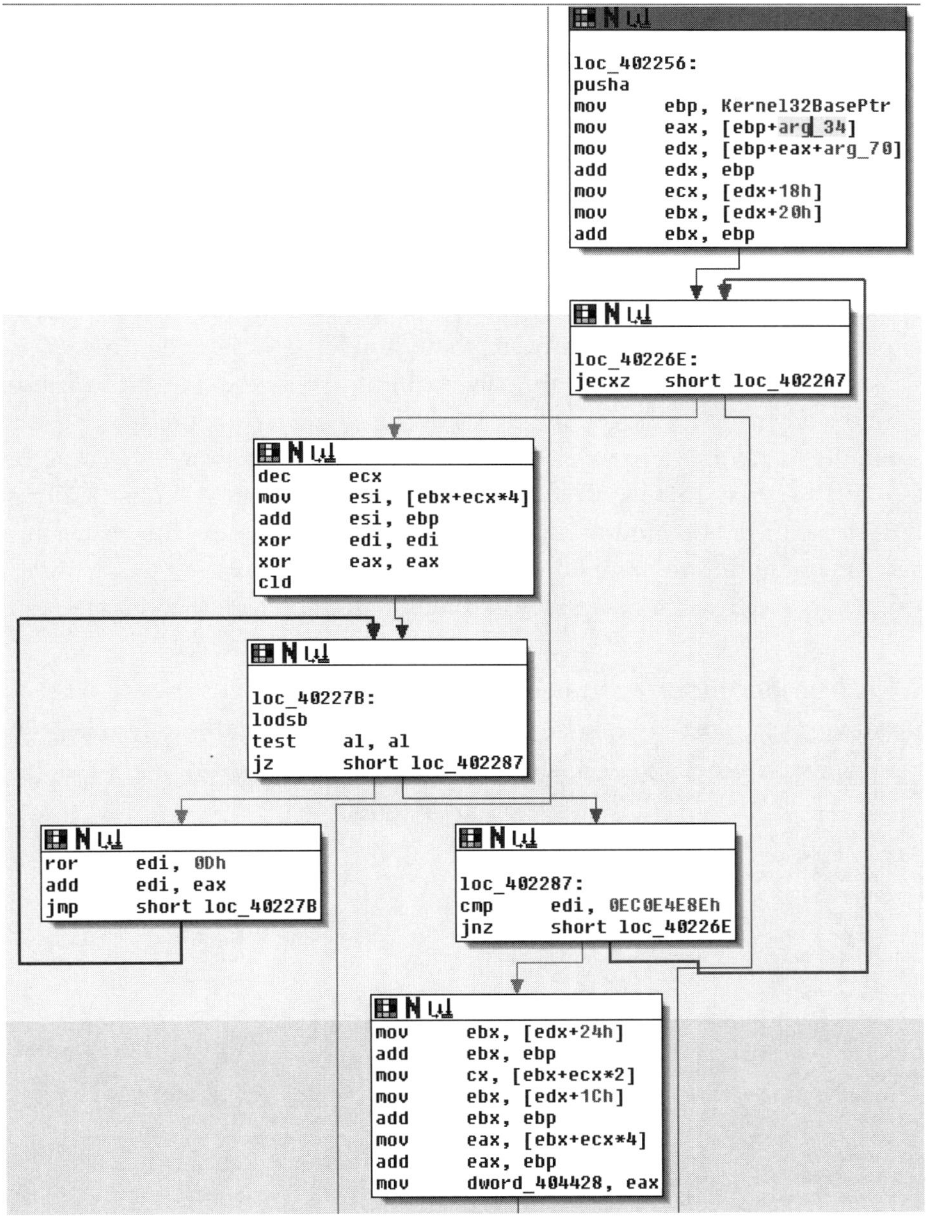

Moving on through the code, at loc_402256, which is where control is handed after the check of the return value from the second call to *FindBaseAddr()* if the base address for kernel32.dll was found successfully, we first see that a *pusha* instruction is executed (Figure 6.13). Some may argue with me on this, but there are instructions like *pusha* that I don't see the compiler generate often and so I often suspect when looking at that code that it may have been hand-written. I know, because I wrote the source in this case, that this is true, but it's an observation that I think is generally true; your mileage may vary of course.

At any rate, we see the Kernel32 base pointer retrieved and it has arg_34 and arg_70 added to the base; we then see offsets 0x18 and 0x20 retrieved from that offset and 0x20 is added to the base pointer. From this point, we could make a guess at what the function does based on what we've already seen and the parameters and offsets being worked with. Plus, if you've done any Windows exploitation, the entire code sequence should look familiar to you, but we'll take a look a little further because something interesting occurred here.

If you look at the first box, loc_402256, you'll notice that arg_34 and arg_70 are used. There are two things that make this odd: the first is that we're inside the *main()* routine, and there was no arg_34 or arg_70. If there were, these should be expressed as offsets from *argv* or *envp*. The second is that these values are only read from, so by looking at this statically we're not even positive what these offsets will be exactly. This is actually a pretty good thing for the person employing anti-reversing; in order for me to really know what's going on in that section of code, I'm going to need to look at it in a debugger and so, unless I skip this part, static analysis stops. In my opinion, what makes this good is that now they can actually take an active role in attempting to complicate the reverser's life, as opposed to a passive/static role that can be achieved when being viewed under a disassembler. (See Figure 6.13).

Figure 6.13 Viewing under a Disassembler

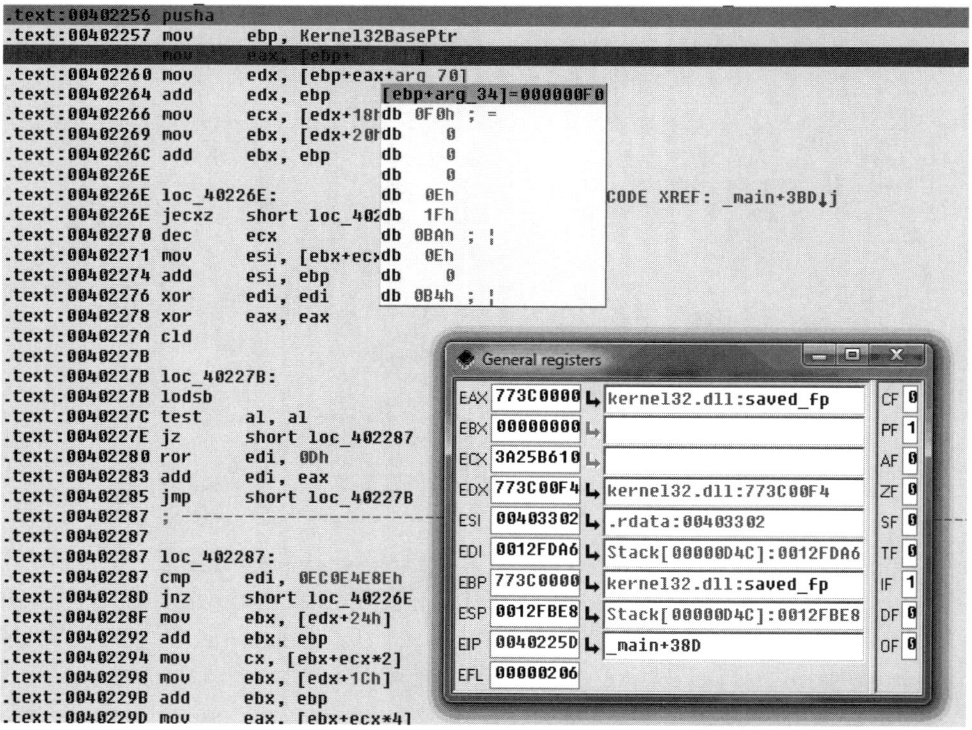

When we look at it in the debugger, we find that this view is not all that much more helpful. However, if we double click on arg_34 and view that memory, what's going on becomes a little more clear (Figure 6.14).

Figure 6.14 arg_34

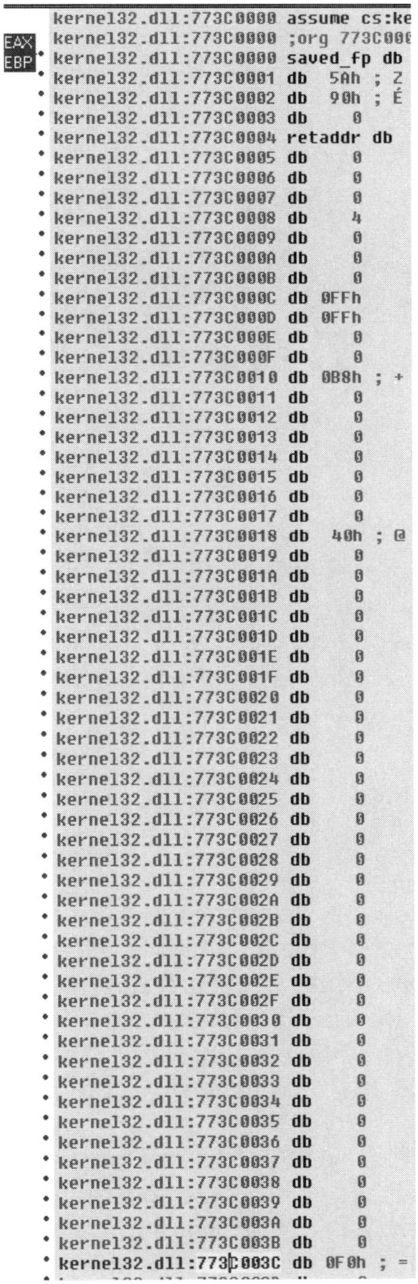

When we view the pointer that IDA is calling arg_34, it becomes clear that it's an offset from the base of Kernel32.dll, specifically the offset 0x3C, which as you may recall from *FindBaseAddr()* is the offset from the base to where the offset to the PE header is specified. This makes sense, a lot more sense than arg_34 implies. Now you may think that this is a great way to obfuscate intent, and indeed it works with limited success and serves mostly as an annoyance. However, it should be noted that other debuggers—such as OllyDBG—may not have the same issues. For instance, Figure 6.15 shows a screenshot from the ImmunitySec debugger, which is a rebranded OllyDBG with Python glued to it.

Figure 6.15 ImmunitySec Debugger

As you can see here, the code is represented correctly in this debugger and its purpose is pretty clear. The main reason this occurs in IDA is because the code fiddles with the EBP register (generally speaking, another tell-tale sign of inline assembly). This in turn causes IDA to confuse the offset and think it's a function parameter. Another reason that could have a noticeable impact, although it is quite likely that fiddling with EBP was enough, is that this was actually a function that was inlined. Either way, after a brief step into the debugger, we know that these two *mov*'s are actually taking offsets 0x3C and 0x78 from the Kernel32 base, respectively. (See Figure 6.16).

TIP

Many people find using IDA's debugger awkward at best and prefer to use one of the many other debuggers available. There are several others. For instance, WinDBG is put out by Microsoft and is a ring-0 debugger, and there is the slowly dying SoftIce which was another ring-0 debugger, and a longtime cracker favorite. However, lack of support for the application and changes in the ways that Windows operates have slowly caused SoftIce to die off due to operability issues.

Another debugger is OllyDBG. This has long been a staple of the reversing communities, largely because it's simple to use, has a fairly intuitive interface and is free. Semirecently, a security firm named Immunity Sec. purchased some form of rights to OllyDBG and combined it with the ability to script Python plug-in's, along with other tweaks, and rereleased it. Immunity Debugger is fairly useful if you're doing exploit development, not only because of scripting capabilities, but also because it ships with useful scripts and features such as identification of heap metadata and so on. If you do much exploit development on Windows platforms and haven't at least tried Immunity's debugger offerings, you really should.

Figure 6.16 Offsets 0x3C and 0x78 from the Kernel32 Base

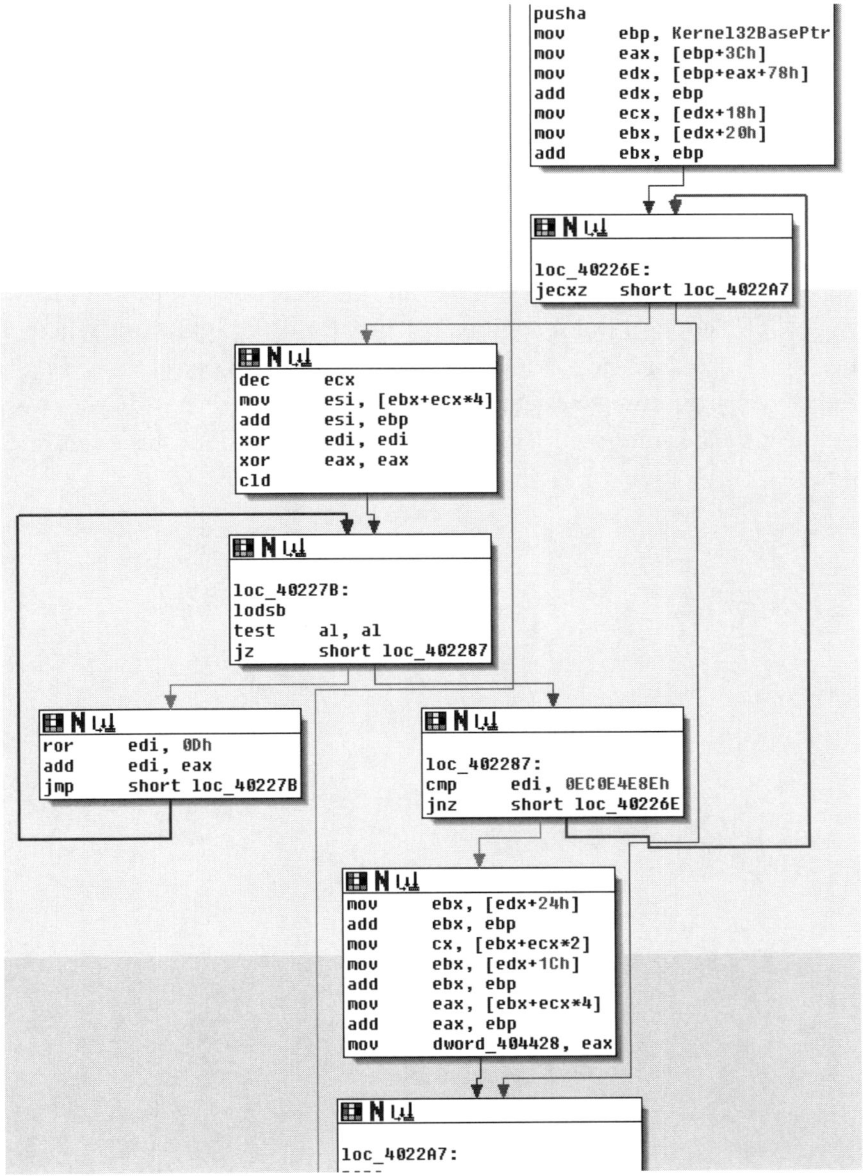

At any rate, fixing the misrepresentation by IDA is easy enough; by right-clicking on the variable name we are prompted with a list of different representations, the first option being the correct one. Thus, we really didn't need to use the debugger. Considering the rest of the code here, this is actually a fairly familiar sequence of code that, as far as the author is aware of, employs a technique first publicly divulged by a Polish hacker group named the Last Stage of Delirium (LSD) and then later expanded upon and reiterated upon in a paper by Skape and the nologin crew, "Understanding Win32 Shellcode." What we see is that the PE header is found at offset 0x3C, then this offset plus the base plus 0x78 yields the Exports data directory. The rest of the code is simply iterating over Exports and taking the names and hashing them with a ror instruction. This result is then compared with another 4-byte hash and, if it matches the export in question, has been found.

In other words, this is just a position independent way of finding a DLL's export without depending on any other APIs. This is often used during the "bootstrapping" process of shellcode. It's typically used to find the address of functions like *LoadLibrary()* and *GetProcAddress()*. Consequently, if you look at loc_402287 you will see a cmp of EDI with the constant value 0x0EC0E4E8E. This is the hash that this code is looking for and if you simply pop it into Google (or run it through a debugger) you would find that this is the hash that corresponds to *LoadLibraryA()*. The code in the second to last box from the bottom is where a match was found, and thus we can rename the variable *dword_404428* to be a pointer to *LoadLibraryA()* and move on. I didn't really dig into the details of the algorithm here, mostly due to space requirements, but if you're interested I strongly advise that you look up either the paper by Skape or LSD.

Figure 6.17 Kernel32

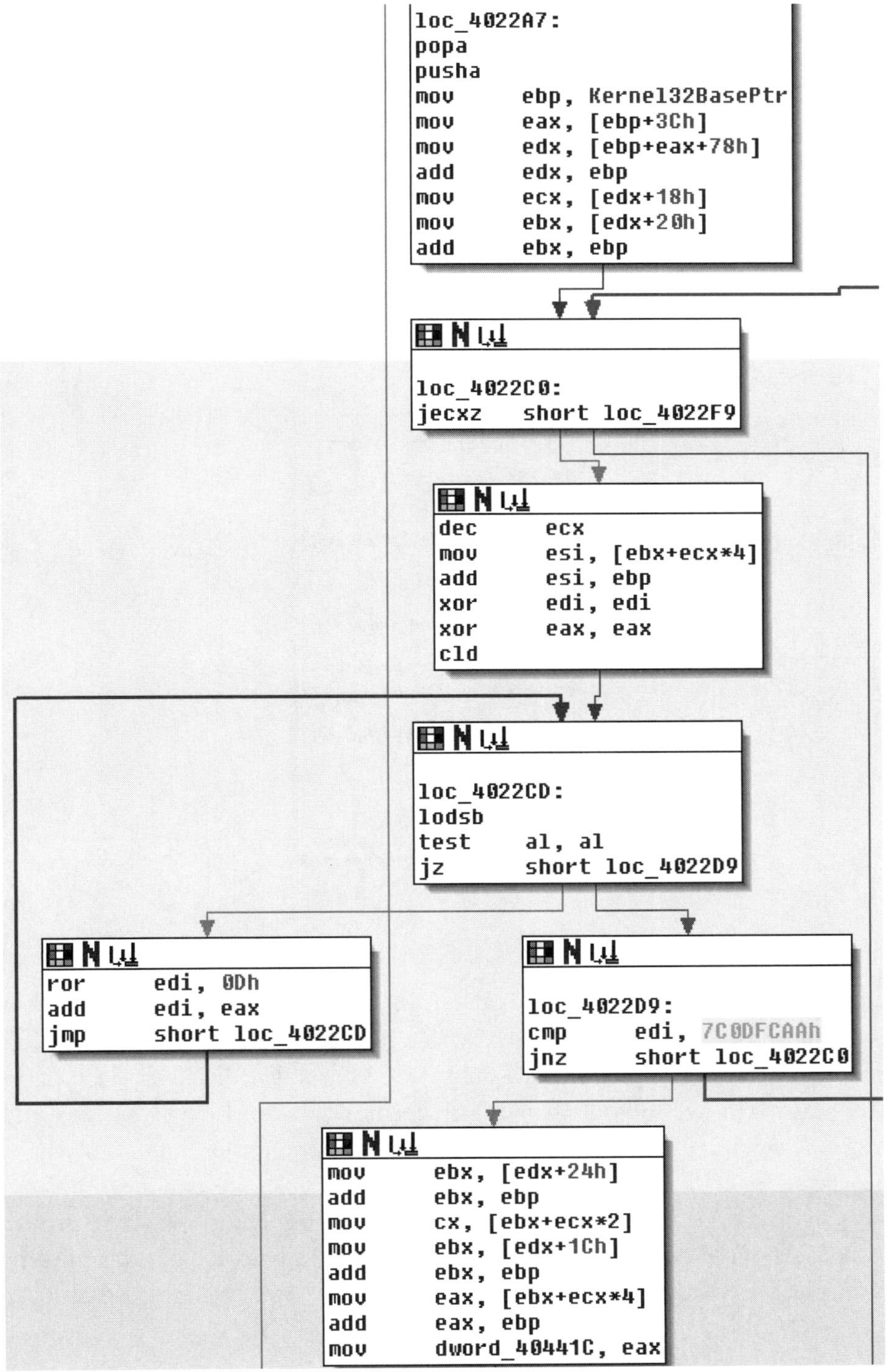

Moving along, we notice a startling familiarity in the next section of code (Figure 6.17), almost to the point that you may wonder if I reposted the wrong image! I assure you I did not. What we see here is another walk through Kernel32 in the same manner in an attempt to find another exported function. This time the hash in question can be found at loc_4022D9 and has a value of 0x7C0DFCAA. Once again, either via Google or a debugger, you would find pretty quickly that this is the hash value for the *GetProcAddress()* function; thus this section of code just locates that pointer, and saves it at *dword_40441C*, which we will rename to *GetProcAddressPtr*. One other thing the reader might note is that this section of code did not exhibit the same oddity as earlier when extracting offsets 0x3C and 0x78. This is simply the result of my changing its representation already.

Figure 6.18 loc_4022F9

```
loc_4022F9:
popa
push    0Ah
lea     eax, [ebp+var_6C]
push    eax
call    sub_4014D0
add     esp, 8
lea     ecx, [ebp+var_6C]
push    ecx
call    LoadLibraryAPtr
mov     dword_40442C, eax
push    0Ah
lea     edx, [ebp+var_6C]
push    edx
call    sub_4014D0
add     esp, 8
cmp     dword_40442C, 0
jnz     short loc_402345
```

Immediately following the previous code, in Figure 6.18, we find loc_4022F9 in which we can see a call to another new subroutine whose return value is stored in *var_6C*. Following that we see a call to *LoadLibraryA()* so it's a pretty good guess that this likely decrypts or decodes some of that stack data we looked at first. Furthermore, we find that we'll save the return value from *LoadLibraryA()* in *dword_40442C*. Finally, we get another clue as to what to expect for the encryption/encoding, as after the usage of the string we see that a call to *sub_4014D0* is made again with the same argument of *var_6C*, at least implying that we're probably going to be looking at a symmetric cipher of some sort. We'll take a brief look at this implementation just to get an idea of what's going on in there, but once you've confirmed that it's just some sort of string obfuscation scheme that encrypts/decrypts itself, it's typically fastest to simply let it do its thing and copy the results out.

That said, you should probably at least read the source of the function you're going to step over to ensure it does what you think it does; we'll largely leave this as an exercise for the reader however, as the book gains little from a detailed analysis of the crypto employed and it would take up a significant amount of space in a chapter that's already tight. Plus, it's trivial but not trite, so it will work out as a good educational exercise for the inspired reader.

Figure 6.19 Reloading the Debugger

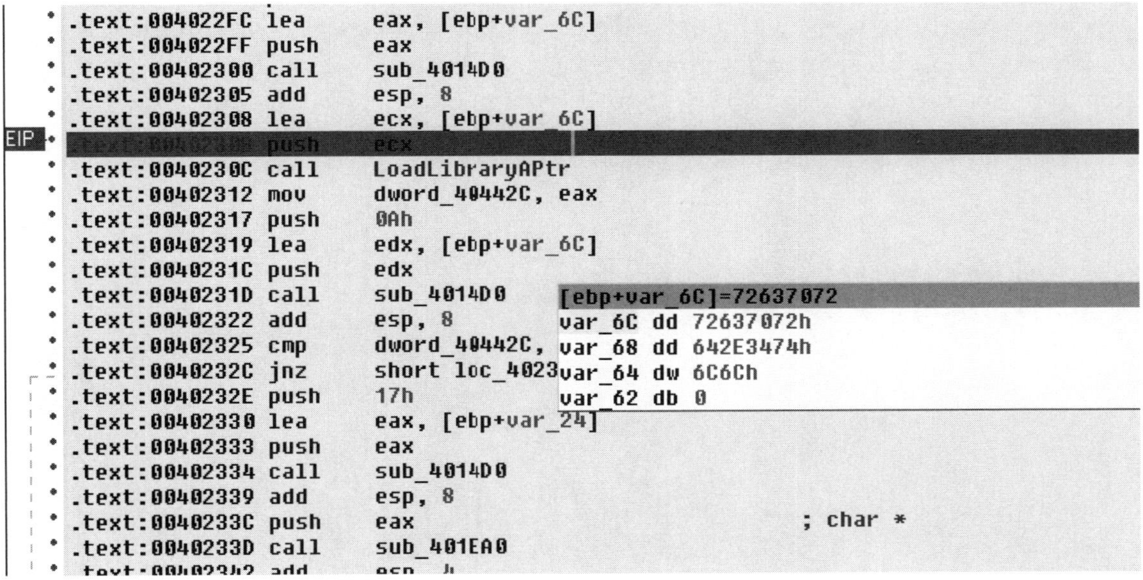

Reloading the debugger, letting the program do the hard work and decrypt the string for us, what we see post calling *sub_4014D0* is the hex sequence seen in Figure 6.19. If you look it up, this corresponds to the ASCII mapping for the NULL terminated string "rpcrt4.dll"—the library required to make RPC calls in Windows. This makes sense, of course, given that we know this to be an RPC server from our earlier analysis. This also largely pieces the puzzle together, as we can likely guess that the purpose of the *GetProcAddress()* pointer. Once again, to conserve space and leave something to do for the reader, we'll leave the rest as an exercise.

Summary

To review the techniques described in this chapter, we've seen the code base make-up change drastically and the complexity of the program increase rapidly as well. We've seen how easy it is to remove the string constants and similar. The interested reader who finds this particular section interesting might enjoy reading more about things like overwriting pointers in the Import Address Table (IAT), or copying system functions into user allocated space to avoid breakpoints on functions. Another technique that originated from the virus-writing world is a technique called *entry point obscuring* or EPO. Traditional viruses would modify the entry point in the executable header and typically append themselves to the executable file. This of course yielded a tell-tale sign of infection and gave anti-virus an easy target. As a result, EPO viruses started to appear. EPO viruses, instead of modifying an entry point, will scan the executable section for a jmp or similar and modify that to hand control to the virus elsewhere. This same technique could be used to take advantage of the fact that IA-32 machines have limited debugging support by entering into system libraries 5 to 10 bytes into the function instead of at its original entry point.

Chapter 7

Walkthrough Four

Solutions in this chapter:

- Tracing Execution Flow from a Read Event
- Determining the Structure of a Protocol
- Determining if the Protocol has any Undocumented Messages
- Use IDA to Determine the Functions that Process a Particular Message

The Protocol Problem

It's not uncommon to be presented with an executable where the protocol is either partially unknown or completely unknown. As a reverse engineer, it's your job to either figure out the protocol for compatibility or to check a program for any hidden features that may cause security problems. In this chapter we'll cover tracking a protocol through a binary and recovering its message structure.

Protocol Structure

Most protocols are streams of discrete messages meant to be interpreted individually. There are exceptions to this rule. HTTP, for instance, dumps off a mostly unstructured request and then gives an unstructured reply. FTP uses a text-based control channel and establishes a separate TCP session for each file transferred. These are in the minority of protocols you'll have to reverse.

If a reverse engineer doesn't have access to either of the executables, the protocol can be reversed from only the raw bytes on the wire. If a reverse engineer only has access to the binary without the ability to run it, the protocol can eventually be extracted from the executables. In most cases, the engineer will have access to both the binaries and a working implementation. A hybrid approach is the fastest way to solve the problem. The bytes on the wire give the reverse engineer a quick picture of how the protocol is structured, but, in the end, any features not exercised by the client or server will have to be extracted from the binary. If the reverse engineer has the time, the binary always gives the most accurate view.

Framing and Reassembly

Every protocol needs to know where one message ends and the following message begins. This is commonly referred to as *framing*. Most protocols can survive getting out of synchronization with the other end of the connection. If the sender thought a message was 30 bytes long and the receiver interpreted the message as 20 bytes long, the receiver might try to interpret the remaining 10 bytes as a new message, resulting in a corrupt message and the connection being dropped.

TCP/IP doesn't guarantee a message will remain together as it makes its way across the Internet. The message may be broken up into smaller pieces along the way. The mechanics are unimportant to the programmer. What *is* important is that a single call to a *read()* function may return a whole message or some piece of the message. It might also include more than one message. It's up to the application to make sure it has read a full message into the buffer before continuing. I'd like to be able to say that most programs do this well. Unfortunately, most programs you'll encounter in the wild do a very poor job of reassembling and then parsing the messages.

Most small or quickly written programs will assume that messages are transmitted whole across the Internet. Their basic block diagram looks something like Figure 7.1.

Figure 7.1 Basic Block Diagram of a Small Program

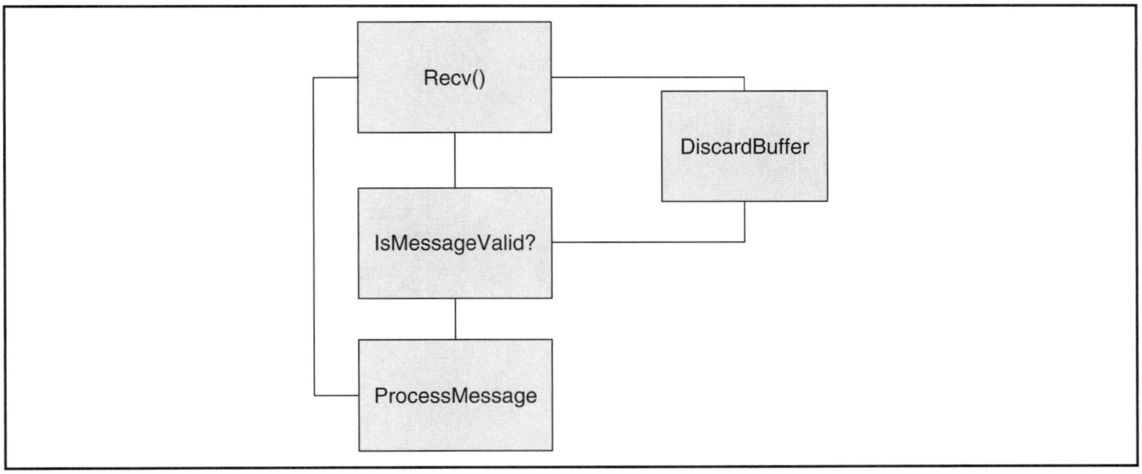

If a message is broken up en route, the program will either crash while parsing the message or reject the message as incomplete, depending on how the function was implemented. Programs written this way still work surprisingly well. When a message is fragmented, the message is returned to the program broken across two reads from the socket. Both the messages are discarded as invalid, but the next message (if it wasn't also broken up) will be received and processed correctly by the system.

A popular instant messenger client is among the programs that implement this kind of loop. Spotting these programs is easy. Simply bring up a proxy between the client and the server and break the messages up in small pieces. If the messages are ignored, you have a program that assumes the messages will be delivered as a unit.

Larger commercial programs often buffer reads into a read buffer and then shift the read buffer down as full messages are received. Their basic block diagram looks like Figure 7.2.

This is a perfectly correct loop and easy to trace through the system. The receive buffer tends to be allocated as a global or on the heap, and if it's properly implemented works in all cases.

Figure 7.2 Basic Block Diagram of a Large Commercial Program

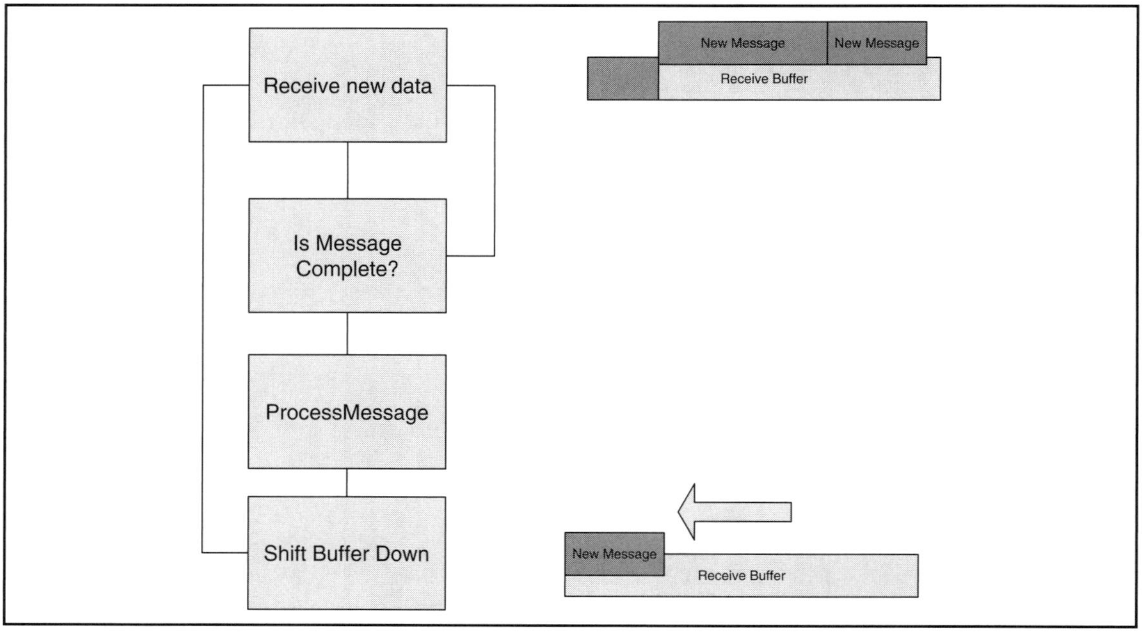

Self Similarity

Protocols are self-similar, meaning that since operations are repeated over the life of the session, the bytes transferred over the wire also repeat. Nearly every protocol has a common protocol header. That header will appear in every message on the wire. At a minimum, the header will contain a length and a message type. The parser needs to know how big the current message is and how to parse it. The following list shows some of the fields often found in the base protocol header.

- Magic Number
- Sequence Number
- Timestamp
- Data or Section Lengths
- Session ID
- Number of Submessages
- Error Code
- Random Nonce

When starting to make sense of a protocol, it helps to have examples of the various packets in front of you for reference. I always print out the hex dumps so I can annotate them as I go. Below are the first few packets from the example used in this section. I've highlighted the packets going from the client to server in **bold** and the packets going from the server to the client in *italic*.

DE AD BE EF 00 18 01 00 76 B7 0B 5A 42 DD 54 B9 6B E8 1E 47 44 D9 67 C3

DE AD BE EF 00 18 02 00 C1 06 45 18 90 51 5D 71 44 46 D7 21 B6 4C 01 73

DE AD BE EF 00 18 01 00 45 EB 2E 08 42 68 22 72 60 E9 1B 32 16 FA 45 EB

DE AD BE EF 00 18 02 00 2E 08 42 68 22 72 60 E9 1B 32 16 FA 40 AB 55 AD

DE AD BE EF 00 18 01 00 64 D4 7F 88 7E E1 5A AA 21 46 49 3D E3 22 7E 1E

DE AD BE EF 00 18 02 00 79 18 D2 6C D7 3D C9 61 60 7B 02 00 DC 4F 40 59

DE AD BE EF 02 08 03 01 AF 59 3E 31 ED 45 FD 02 E3 26 A1 1B 08 12 19 05 16 82 59 26 3B 90 77

DE AD BE EF 02 08 03 00 77 0D 33 A4 03 19 4D F1 62 F5 1F B2 20 CB 37 82 25 87 46 0E 6E 8A 56

DE AD BE EF 02 08 04 00 77 0D 33 A4 03 19 4D F1 62 F5 1F B2 20 CB 37 82 25 87 46 0E 6E 8A 56

DE AD BE EF 02 08 04 00 AF 59 3E 31 ED 45 FD 02 E3 26 A1 1B 08 12 19 05 16 82 59 26 3B 90 77

DE AD BE EF 00 08 01 02 00 00 00 00

At this point, we have no idea how many different types of messages are in this dump and we have no idea how long the base header is. Looking at it column by column, we can start making some guesses. The first four bytes have the same value in every packet. Since they spell DEADBEEF, we can safely assume that they're a magic number at the beginning of the protocol. The shortest packet in this dump is 12 bytes long. It's safe to say that the base header is smaller than the smallest message observed. That leaves us with the start of a structure that looks like this:

```
struct base_header{
       int MagicNumber;      /*Always 0xDEADBEEF*/
       char unknown1;        /*00,02*/
       char unknown2;        /*08, 18*/
       char unknown3;        /*01, 02, 03, 04*/
       char unknown4;        /*00,01,02*/
       char unknown5;        /*Lots of possibilities*/
       char unknown6;        /*Lots of possibilities*/
       char unknown7;        /*Lots of possibilities*/
       char unknown8;        /*Lots of possibilities*/
}
```

The packets have to be read off the network and into the program. There are a limited number of API calls that accomplish the task. The program can directly make the system calls itself, but that's only encountered in malware. The place(s) where the packet is read off the wire is always a good place to start your analysis. The following is a list of common API calls capable of reading traffic off a network socket.

- read/write
- recv/send
- recvfrom,/sendto
- WSARecv/WSASend
- WSARecvFrom/WSASendTo
- ioctl
- ioctlsocket
- WSARecvDisconnect/WSASendDisconnect
- WSARecvEx/WSASendEx
- recvmsg/sendmsg
- WSARecvMsg/WSASendMsg

Figure 7.3 Import Table of Executable Showing Reference to WSARecv

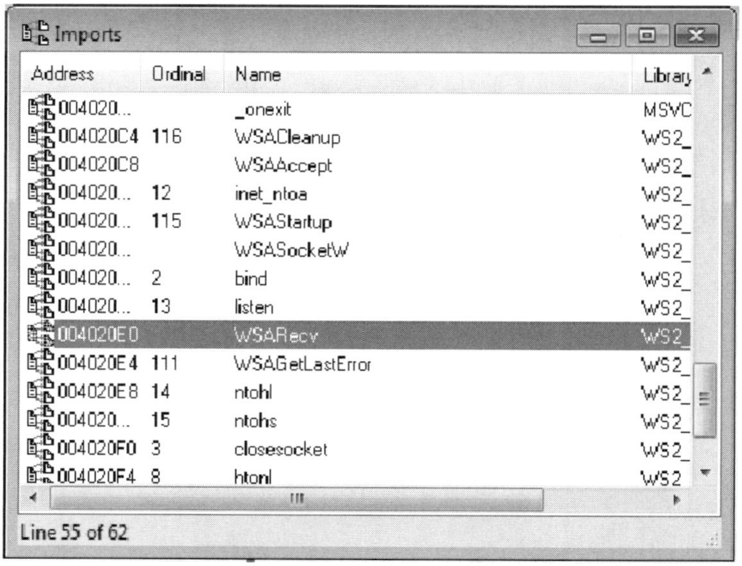

The import table of the executable only has a reference to WSARecv so that serves as in ideal starting point for the analysis of the protocol (see Figure 7.3). The WSARecv call reads the data into a buffer on the stack. It can read up to 0x4000 bytes in one read as shown in Figure 7.4.

Figure 7.4 WSARecv Call

```
push    4000h               ; size_t
push    0                   ; int
lea     ecx, [ebp+var_4018]
push    ecx                 ; void *
call    memset
add     esp, 0Ch
mov     [ebp+Buffers.len], 4000h
lea     edx, [ebp+var_4018]
mov     [ebp+Buffers.buf], edx
push    0                   ; lpCompletionRoutine
push    0                   ; lpOverlapped
lea     eax, [ebp+Flags]
push    eax                 ; lpFlags
lea     ecx, [ebp+NumberOfBytesRecvd]
push    ecx                 ; lpNumberOfBytesRecvd
push    1                   ; dwBufferCount
lea     edx, [ebp+Buffers]
push    edx                 ; lpBuffers
mov     eax, [ebp+s]
push    eax                 ; s
call    ds:WSARecv
mov     [ebp+var_401C], eax
cmp     [ebp+var_401C], 0
jz      short loc_40128C
```

Right after reading into a buffer on the stack, the program immediately checks to see if the number of bytes read is less than eight. If it's fewer than eight bytes, it jumps to the exit function, as shown in Figure 7.5.

Figure 7.5 Exit Function

```
loc_40128C:
cmp     [ebp+NumberOfBytesRecvd], 8
jnb     short loc_4012AA
```

It then checks the first four bytes to see if they're DEADBEEF. That matches the magic number from the dump (the first four bytes) nicely. The bytes read off the wire are passed through *ntohl()* before they're compared. This means the wire protocol is big-endian. It will also help to determine the size of fields. The program will need to byte-swap all two- and four-byte fields before processing them, as shown in Figure 7.6.

If it passes the DEADBEEF test, it takes the next two bytes and checks to see if they're less than eight. Packets with a small number are discarded. We now know that the two bytes

Figure 7.6 Byte-swapping

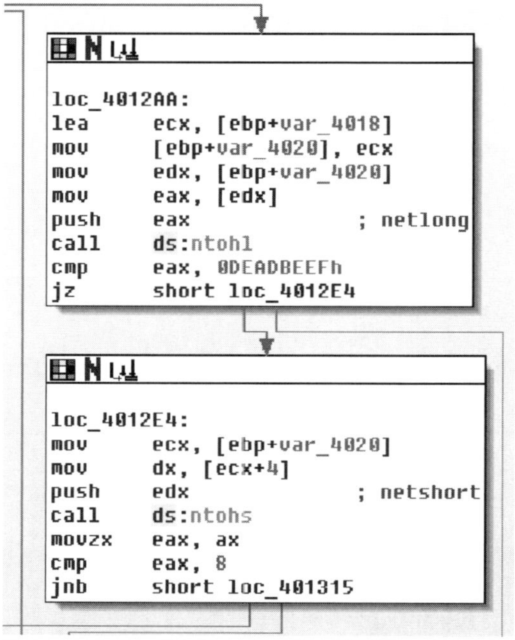

following the magic number are a unit and they always have to be greater than eight. With the information obtained from these two basic blocks, the structure diagram can be updated. The number corresponds with the length of the packet, so for now I'll label the two-byte field as the length field.

```
struct base_header{
        int MagicNumber;         /*Always 0xDEADBEEF*/
        unsigned short Len;      /*0018,0208*/ /*Greater than 8*/
        char unknown3;           /*01, 02, 03, 04*/
        char unknown4;           /*00,01,02*/
        char unknown5;           /*Lots of possibilities*/
        char unknown6;           /*Lots of possibilities*/
        char unknown7;           /*Lots of possibilities*/
        char unknown8;           /*Lots of possibilities*/
}
```

After the length check, the processing function pulls a single byte at offset 7 into a register and uses the *and* instruction against the result. Whenever a field pulled out of the protocol is passed to bitwise *and* or *or* operators, and the operand is a constant, it's a good bet that the target is a bitfield of some sort. In this case, if the third bit is set, the process stops processing the packet. It's safe to say that byte 7 is a bitfield and the constant 0x04 is invalid, as shown in Figure 7.7.

Figure 7.7 Use of *and* Operator

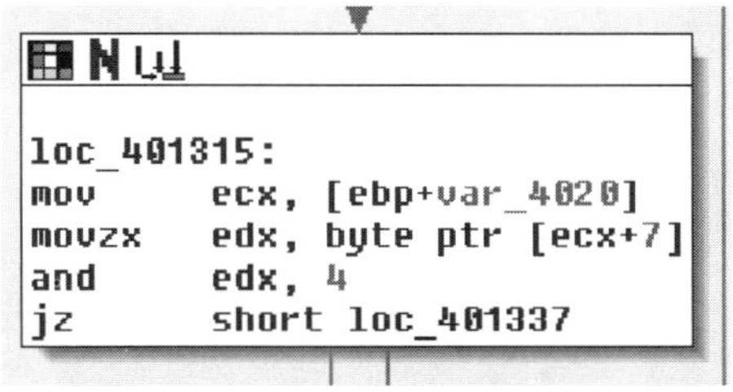

The updated struct looks something like:

```
struct base_header{
       int MagicNumber;          /*Always 0xDEADBEEF*/
       unsigned short Len;       /*0018,0208*/ /*Greater than 8*/
       char unknown3;            /*01, 02, 03, 04*/
#define FLAG_NONE                0x00
#define FLAG_INVALID             0x04
       char Flags;               /*00,01,02*/
       char unknown5;            /*Lots of possibilities*/
       char unknown6;            /*Lots of possibilities*/
       char unknown7;            /*Lots of possibilities*/
       char unknown8;            /*Lots of possibilities*/
}
```

From the packet dump, we saw that two other flags were set. In most flags an empty bitfield means "do nothing," so as a placeholder, zero is defined as FLAG_NONE. That leaves at least 0x01 and 0x02 as valid flags in the protocol.

Next up, the binary takes byte 6 and stores it into a stack variable. It compares the stack variable to three possible values: 0x01, 0x03, and 0x5C. In the dump, we've seen 0x01 through 0x04, but only 0x01 and 0x03 are packets destined for the server. It would seem that 0x02 and 0x04 are client→server constants while 0x01, 0x03, and 0x5C are server→ client constants. All other values aren't processed, as shown in Figure 7.8.

Figure 7.8 Other Values Are Not Processed

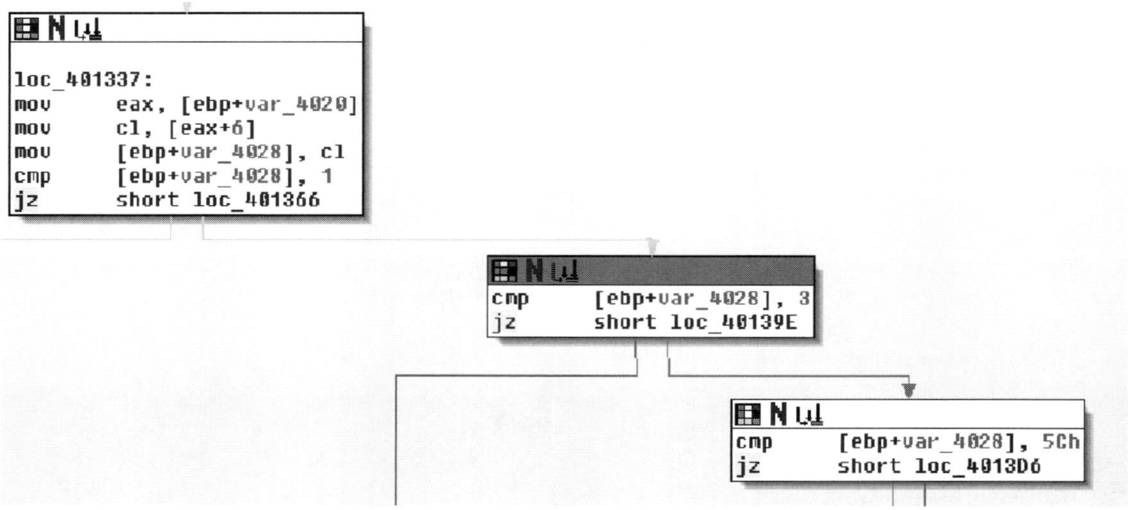

Depending on the value of unknown3, the binary will call one of three functions. The prototype for each of the called functions is the same. The first parameter is a pointer to the raw packet we sent on the wire. The second is the number of bytes read from the network, and the third is the socket handle the packet was read from, as shown in Figure 7.9.

Figure 7.9 Prototype for Each of Three Called Functions

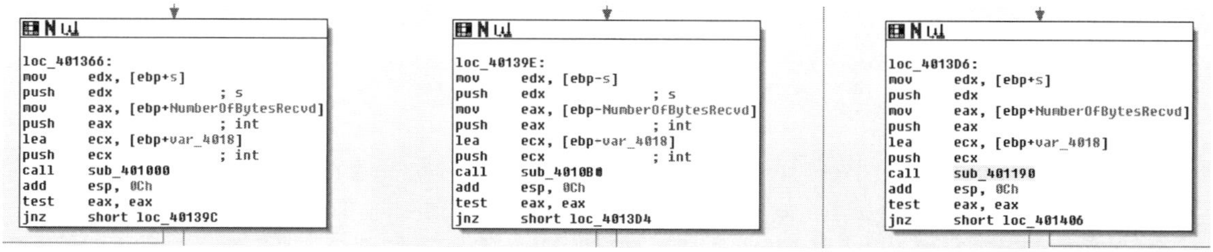

Later in the tutorial, we'll reverse each of the three functions. For now, we'll finish out the logic of the current function. The final two comparisons are both against the flag field. The binary uses a bitwise *and* to check for the 0x01 bit. If the bit is set, the thread goes to sleep for 1000 ms, as shown in Figure 7.10.

Figure 7.10 If Bit Is Set, Thread Goes to Sleep for 1000 ms

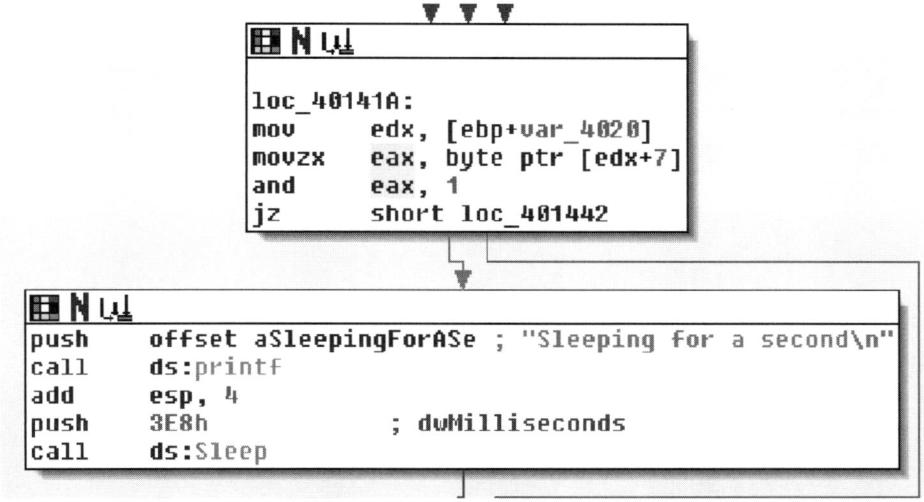

The last check before the end of the processing loop is another bitwise *and* checking for the 0x02 bit in the bitfield. If the bit is set, the processing loop exits, as shown in Figure 7.11.

Figure 7.11 If Bit Is Set, Processing Loop Exits

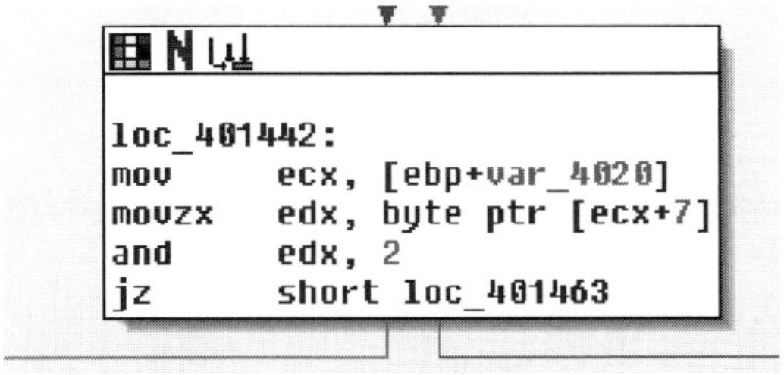

That completes the main parsing loop of the example server. Since we didn't do anything with the other bytes in the struct, it can be safely assumed that they aren't part of the base header in the protocol and those bytes belong to some deeper part of the protocol. The final structure appears in the code below.

```
struct base_header{
       int MagicNumber;            /*Always 0xDEADBEEF*/
       unsigned short Len;         /*0018,0208*/ /*Greater than 8*/
#define PACKET_SERVER_01           0x01
#define PACKET_CLIENT_02           0x02
#define PACKET_SERVER_03           0x03
```

```
#define PACKET_CLIENT_04          0x04
#define PACKET_SERVER_5C          0x5C
        char PacketType;          /*01, 02, 03, 04*/
#define FLAG_NONE                 0x00
#define FLAG_SLEEP                0x01
#define FLAG_PROCESS_AND_EXIT     0x02
#define FLAG_INVALID              0x04
        char Flags;               /*00,01,02*/
}
```

For this simple protocol, all we're left with is to figure out what each of the three message types are for. The server processes messages of type 0x01, 0x03, and 0x5C. The 0x01 message is as good a place to start as any other, since we're trying to reverse the entire protocol.

The function at 0x00401000 handles messages of type 0x01. Three parameters are passed to it: a pointer to the stack buffer containing the full message, the number of bytes read, and the socket it was read from, as shown in Figure 7.12.

Figure 7.12 Three Parameters Are Passed to the Function

```
push    ebp
mov     ebp, esp
sub     esp, 24h
mov     eax, dword_403000
xor     eax, ebp
mov     [ebp+var_4], eax
push    18h                 ; size_t
push    0                   ; int
lea     eax, [ebp+buf]
push    eax                 ; void *
call    memset
add     esp, 0Ch
push    0DEADBEEFh          ; hostlong
call    ds:htonl
mov     dword ptr [ebp+buf], eax
push    18h                 ; hostshort
call    ds:htons
mov     [ebp+var_18], ax
mov     [ebp+var_15], 0
mov     [ebp+var_16], 2
mov     [ebp+var_24], 0
jmp     short loc_401054
```

The first thing you should notice is that the function never touches the buffer or the length fields passed into it. It jumps straight to constructing another message in a local stack buffer. It first zeros the buffer with memset, then byte-swaps the magic number

(DEADBEEF) and stores it in the buffer. Then it byte-swaps and stores a 0x18 in the length field, a 0x02 in the type field, and a zero in the flags field. Then it fills the next 0x10 bytes with random values, as shown in Figure 7.13.

Figure 7.13 Fills Next 0x10 Bytes with Random Values

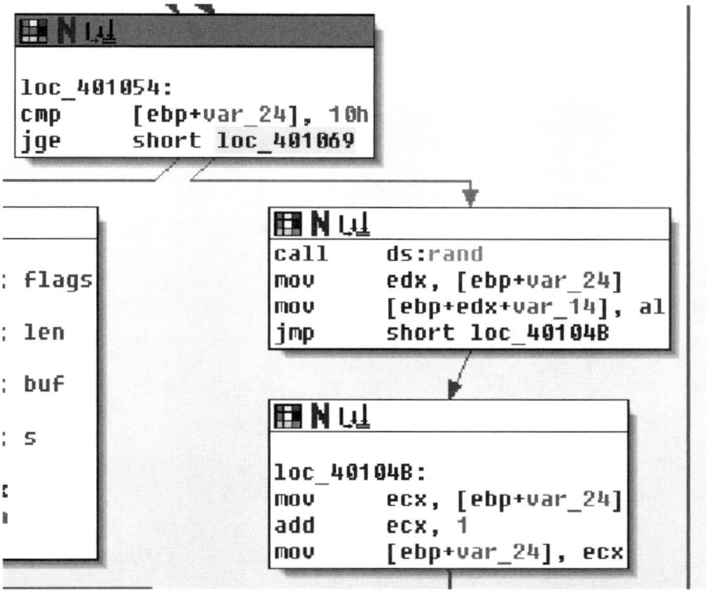

Finally it writes it out to the socket, as shown in Figure 7.14.

Figure 7.14 Writing It Out to Socket

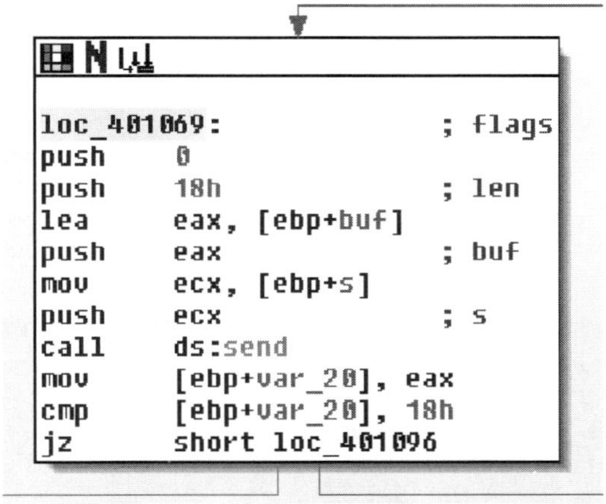

Basically, this generates a predetermined packet and returns it to the caller. The contents of the message received aren't taken into account. We now know the bytes after the header in the calling packet are ignored and the bytes following the 0x02 message are just random values.

```
struct base_header{
        int MagicNumber;                /*Always 0xDEADBEEF*/
        unsigned short Len;             /*0018,0208*/ /*Greater than 8*/
#define PACKET_GETRANDOM                0x01
#define PACKET_RANDOM                   0x02
#define PACKET_SERVER_03                0x03
#define PACKET_CLIENT_04                0x04
#define PACKET_SERVER_5C                0x5C
        char PacketType;                /*01, 02, 03, 04*/
#define FLAG_NONE                       0x00
#define FLAG_SLEEP                      0x01
#define FLAG_PROCESS_AND_EXIT           0x02
#define FLAG_INVALID                    0x04
        char Flags;                     /*00,01,02*/
}
struct packet_get_random{
        struct base_header BaseHeader;
        char Ignored[16];
}
struct packet_random{
        struct base_header BaseHeader;
        char RandomValues[16];
}
```

The next message is processed by the function at 0x004010B0. It's passed the same three parameters as the packet above: a pointer to the message buffer, the message buffer length, and the socket the message was read from.

The first thing it does is to make sure the message length is at least 520 bytes, as shown in Figure 7.15.

Figure 7.15 Ensure Message Length at Least 520 Bytes

```
push    ebp
mov     ebp, esp
sub     esp, 214h
mov     eax, dword_403000
xor     eax, ebp
mov     [ebp+var_8], eax
cmp     [ebp+ReadBufferLen], 520
jnb     short loc_4010D3
```

The entire processing of this packet is done in a single basic block, as shown in Figure 7.16.

Figure 7.16 Single Basic Block Processes Packet

```
loc_4010D3:
mov     eax, [ebp+ReadBuffer]
mov     [ebp+var_214], eax
push    208h            ; size_t
push    0               ; int
lea     ecx, [ebp+buf]
push    ecx             ; void *
call    memset
add     esp, 0Ch
push    0DEADBEEFh      ; hostlong
call    ds:htonl
mov     dword ptr [ebp+buf], eax
push    208h            ; hostshort
call    ds:htons
mov     [ebp+var_20C], ax
mov     [ebp+var_209], 0
mov     [ebp+var_20A], 4
push    200h            ; size_t
mov     edx, [ebp+var_214]
add     edx, 8
push    edx             ; void *
lea     eax, [ebp+var_208]
push    eax             ; void *
call    memcpy
add     esp, 0Ch
push    0               ; flags
push    208h            ; len
lea     ecx, [ebp+buf]
push    ecx             ; buf
mov     edx, [ebp+s]
push    edx             ; s
call    ds:send
mov     [ebp+var_4], eax
cmp     [ebp+var_4], 208h
jz      short loc_401177
```

The basic block zeros a stack buffer with memset. Then it byte-swaps the magic number (DEADBEEF) and stores it in the first four bytes of the buffer. It sets the byte-swapped length to 0x208, the type to 4 and the flags to zero. Then it copies 512 bytes from the incoming packet and puts it in the outgoing packet.

```
struct base_header{
       int MagicNumber;              /*Always 0xDEADBEEF*/
       unsigned short Len;           /*0018,0208*/ /*Greater than 8*/
#define PACKET_GETRANDOM             0x01
#define PACKET_RANDOM                0x02
#define PACKET_ECHO                  0x03
#define PACKET_ECHOREPLY             0x04
#define PACKET_SERVER_5C             0x5C
       char PacketType;              /*01, 02, 03, 04, 0x5C*/
#define FLAG_NONE                    0x00
#define FLAG_SLEEP                   0x01
#define FLAG_PROCESS_AND_EXIT        0x02
#define FLAG_INVALID                 0x04
       char Flags;                   /*00,01,02*/
}
/*must be 520 bytes long or longer*/
struct packet_get_echo{
       struct base_header BaseHeader;
       char RandomData[512];
}
struct packet_echoreply{
       struct base_header BaseHeader;
       char EchoData[512];
}
```

The last message (type 0x5C) is handled by the function at 0x00401190. It's passed the same three parameters as the other message handlers: a pointer to the message, the message length, and the socket the message came in on. The message is handled in a single basic block, as shown in Figure 7.17.

It looks like this message spawns a calculator. The final protocol description is given below.

```
struct base_header{
       int MagicNumber;          /*Always 0xDEADBEEF*/
       unsigned short Len;       /*0018,0208*/ /*Greater than 8*/
#define PACKET_GETRANDOM         0x01
#define PACKET_RANDOM            0x02
#define PACKET_ECHO              0x03
#define PACKET_ECHOREPLY         0x04
#define PACKET_CALC              0x5C
       char PacketType;          /*01, 02, 03, 04, 0x5C*/
```

```
#define FLAG_NONE                0x00
#define FLAG_SLEEP               0x01
#define FLAG_PROCESS_AND_EXIT    0x02
#define FLAG_INVALID             0x04
        char Flags;              /*00,01,02*/
}
struct packet_get_random{
        struct base_header BaseHeader;
        char Ignored[16];
}
struct packet_random{
        struct base_header BaseHeader;
        char RandomValues[16];
}
/*must be 520 bytes long or longer*/
struct packet_get_echo{
        struct base_header BaseHeader;
        char RandomData[512];
}
struct packet_echoreply{
        struct base_header BaseHeader;
        char EchoData[512];
}
struct packet_calc{
        struct base_header BaseHeader;
}
```

The protocol description above should be enough to implement a client. We're now also sure that this is all of the functionality of the server—there are no hidden uses the server can be put to.

Hit Marking

The example used in the first part of this section is a simple one. There are only a few functions and all of them directly deal with the protocol. Reverse engineering the entire program is feasible for the example, but it's rare that a reverse engineer will have the luxury of reversing an entire program. It's easy to find the functions to tear apart when a program's call tree looks like Figure 7.18.

It's much more difficult when the call tree looks like Figure 7.19.

The first question the reverse engineer must answer is which of the functions in this monstrous graph are used in processing messages and which ones are ignored. The most

Figure 7.17 Single Basic Block That Handles Last Message

```
; Attributes: bp-based frame

; int __cdecl sub_401190(char *ReadBuffer,int ReadBufferLen,SOCKET s)
sub_401190 proc near

var_4= dword ptr -4
ReadBuffer= dword ptr  8
ReadBufferLen= dword ptr  0Ch
s= dword ptr  10h

push    ebp
mov     ebp, esp
push    ecx
push    1                ; uCmdShow
push    offset CmdLine   ; "c:\\windows\\system32\\calc.exe"
call    ds:WinExec
mov     [ebp+var_4], eax
mov     eax, 20h
cmp     eax, [ebp+var_4]
sbb     eax, eax
neg     eax
mov     esp, ebp
pop     ebp
retn
sub_401190 endp
```

Figure 7.18 Easily Identified Functions in a Program Call Tree

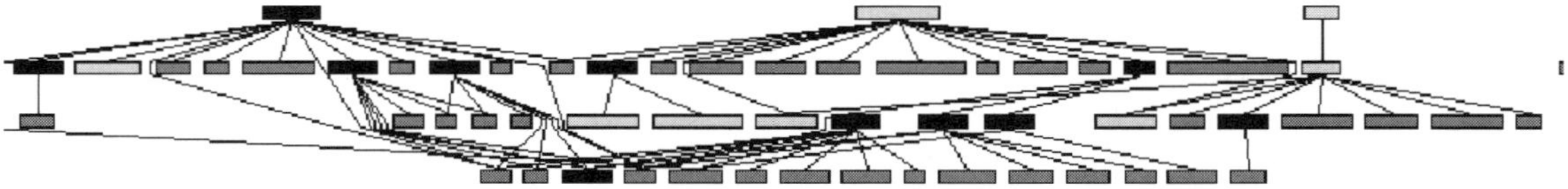

Figure 7.19 More Difficult to Identify Functions in a Program Call Tree

common method of picking which functions to reverse engineer is referred to as *hit marking*.

The basic concept behind hit marking is to record the path through the decision tree that each message takes. A breakpoint is set at the beginning of each function or basic block. When a breakpoint is reached, the program notes that the function was accessed and lets the program continue on to the next function. After the message has been fully processed, the list of breakpoints that were hit is sorted and duplicates are removed. This list is referred to as a *hit list*. By analyzing just the functions on the hit list, the reverse engineer should have an understanding of the structures that make up that particular message.

Figure 7.20 shows a short trace from IDA before duplicates are removed.

Figure 7.20 Short Trace from IDA

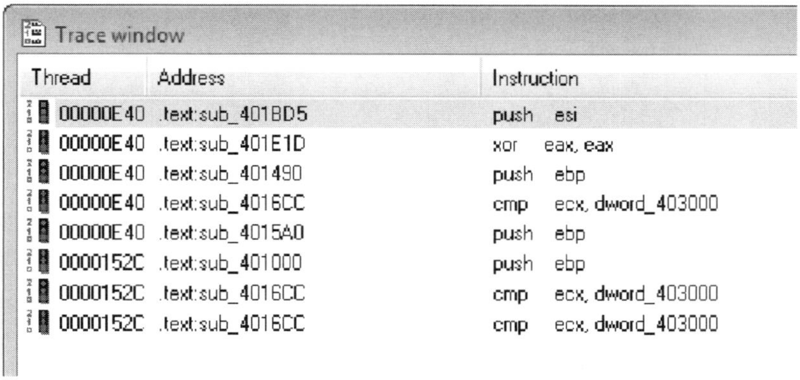

Unfortunately, IDA doesn't have a good mechanism for building hit lists. A number of third-party plug-ins and applications can be purchased to make the process more useful and less painful, but we're going to stick to pure IDA. The rest of this chapter will walk through developing a hit list.

The first thing you'll need to build a hit list is a list of all the functions in the program. If you happen to have a way of removing functions that you know you're not interested in, then by all means remove them. To obtain a list of functions, open your binary in IDA and open the function window (**shift-f3**). Right click and select **copy** (or hit **ctrl-ins**). IDA will copy the list of functions to the clipboard. Paste the list of functions into your favorite editor. You'll get a list something like:

```
_IID_ISAXErrorHandler      .text 01001308 00000010 R . . . . . .
_IID_IXMLDOMDocument2      .text 01001318 00000019 R . . . . T .
_IID_ISchemaElement        .text 01001338 00000009 R . . . . . .
```

```
_IID_ISchemaAttribute       .text 01001348 00000009 R . . . . . .
_IID_ISchemaModelGroup      .text 01001358 0000000D R . . . . . .
_IID_ISchemaComplexType     .text 01001368 0000000D R . . . . . .
_IID_ISchemaType            .text 01001378 0000000D R . . . . . .
_IID_ISchemaItem            .text 01001388 00000009 R . . . . . .
_IID_ISAXAttributes         .text 010013A8 00000010 R . . . . . .
```

Use your text edit to remove everything but the function addresses. It should look something like:

```
01001308
01001318
01001338
01001348
01001358
01001368
01001378
01001388
010013A8
```

The IDA debugger has a built-in mechanism for tracing. It can be accessed from the **Debug | tracing window** or you can right-click on any instruction and add an execution trace. Theoretically, you could manually run through your entire function list and manually add a trace to each breakpoint. Four days later, your boss will probably fire you.

To efficiently add tracepoints to each function, we need to write a quick IDA plug-in. IDA plug-ins can be written in C, ruby, or python. Programming language holy wars aren't a good thing, so I'm going to sidestep the issue and just write the plug-in in python because I wanted to (not because python is better). You can get the python plug-in from www.d-dome.net/idapython/. Just follow the instructions to install it.

What we want the plug-in to do is to automate setting all those hundreds of tracepoints. As a simple example, you can set a breakpoint at 0x004011E8 with just a few lines of python. Create a file with the following content:

```
#Set a breakpoint at 0x004011E8
from idautils import *
ea=ScreenEA()
ea=0x004011E8
add_bpt(ea, 1, 4)
```

Now run the file by hitting **alt-9** and selecting it from the dialog box. (The file has to end in .py if you're using the python bindings.) A breakpoint should now be set at 0x004011E8. You can check to see if it worked by looking at the breakpoints window in Figure 7.21. The hotkey for the breakpoints window is **ctrl-alt-B**.

Figure 7.21 Breakpoints Window

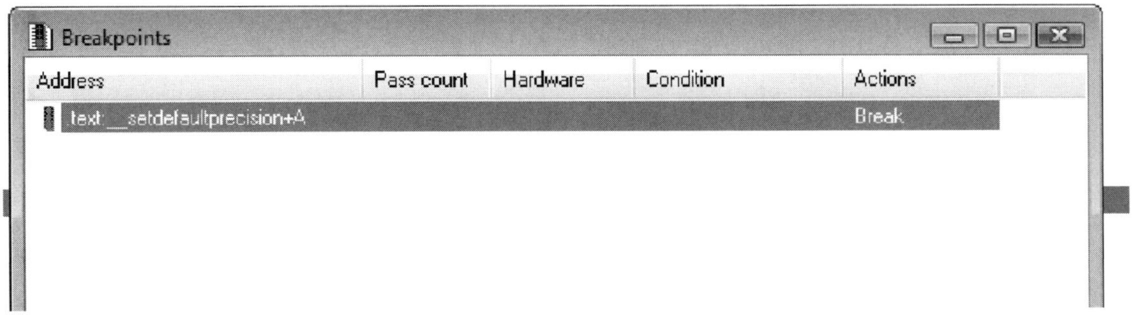

The new breakpoint is set to break. What we want is for it to create a log entry every time the breakpoint is hit. In IDA terminology, that's a trace instead of a break. Actually, tracing and breaking aren't mutually exclusive. You can set it to stop execution (break) when it hits the breakpoint and log the event (trace) at the same time. Since we're just looking for which functions are executed when processing a particular message, we want all the breakpoints set to trace. We have to remove the break flag and add the trace flag from the structure. The following code adds a breakpoint at 0x004011E8 and sets it to trace.

```
from idautils import *
ea=ScreenEA()
ea=0x004011E8
add_bpt(ea, 1, 4)
bp=bpt_t()
get_bpt(ea, bp)
bp.flags^=BPT_BRK
bp.flags|=BPT_TRACE
```

Now that we can set a tracepoint anywhere in IDA, with the list of function addresses from before, it's a simple task to add a loop and set a tracepoint at the beginning of each function.

```
from idautils import *
funclist=[0x004011F2, 0x00401200, 0x00401206]   /*add all the other
addresses here*/
ea=ScreenEA()
for i in funclist:
        ea=i
        add_bpt(ea, 1, 4)
        bp=bpt_t()
        get_bpt(ea, bp)
        bp.flags^=BPT_BRK
        bp.flags|=BPT_TRACE
```

For those cases where it's necessary to mark absolutely every function as a tracepoint, the IDA plug-in can be automated to find every function reference instead of doing it manually. In practice, you'll probably have a number of functions you want to eliminate off the bat, so the example above will be used more than the fully automated one. Just for completeness, here's a python snippet that automates setting a tracepoint at every function.

```
from idautils import *
# Loop through all the functions and add a breakpoint
for i in range(get_func_qty()):
    f=getn_func(i)
    print "Function %s at 0x%x" % (GetFunctionName(f.startEA), f.startEA)
    add_bpt(f.startEA, 1, 4)
#change all the breakpoints to trace-only
for i in range(get_bpt_qty()):
    b=bpt_t()
    getn_bpt(i, b)
    b.flags^=BPT_BRK
    b.flags|=BPT_TRACE
    update_bpt(b)
```

Now that there's a tracepoint set at the beginning of each function, simply run the client program to generate a hit list.

Example Hitlist

Small examples don't really get the point across, so a larger program is in order. I'm going to use Pidgin as an example (www.pidgin.im). Pidgin is a popular open-source chat program that supports most of the popular protocols. Since it's open source, you can compare the disassembly against the binary if you get lost.

In this example, I'm going to generate a hit list against Pidgin v2.1.1 for Windows. Pidgin implements each protocol as a plug-in. I just picked the Yahoo! Instant Messenger plug-in at random. The protocol logic is implemented in libyahoo.dll. The call graph of the DLL is large and complex. Graphing it in IDA gives a blob that can't be interpreted by humans, like the graph in Figure 7.22. (Or at least by humans that ever leave the house.)

Figure 7.22 "Blob" Graph from IDA

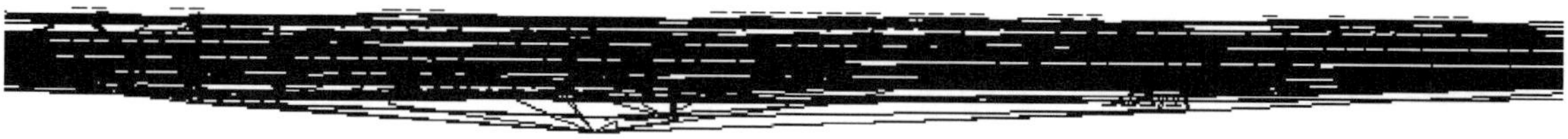

In this example, I'm going to mark every function as a tracepoint with the script in the previous section. There are 549 functions in the DLL. For the first pass, I'm only interested in the functions used during initialization and login. The hit list should narrow the scope down considerably from the 549 functions to something more manageable.

First, create an account on Yahoo! and log onto it with Pidgin to make sure everything is up and working. Next open libyahoo.dll in IDA and set the **Debugger→Process Options** to run Pidgin when the debugger is launched. Finally, invoke the script from the previous section with **alt-9** to mark every function as a tracepoint. Tell IDA to run and go get a cup of coffee. Running the executable under IDA's debugger takes considerably longer than running the executable without the debugger. When the bar at the bottom of the window says **Available**, Pidgin has finished logging in, as shown in Figure 7.23.

Figure 7.23 Logging In Complete

Now we can take a look at the trace window and determine which functions were involved in connecting to Yahoo!. At most we'll have to reverse engineer this set of functions to figure out the login portion of the protocol so it gives us a good upper bound on the amount of work ahead of us.

The trace window (**Debugger | Tracing | Trace Window**) lists the functions in the order they were called. The breakpoint list (**Debugger | Breakpoints | Breakpoint List**) is closer to a true hit list. It lists each breakpoint and a count of the number of times that breakpoint was invoked, as shown in Figure 7.24.

Figure 7.24 Invoked Breakpoints

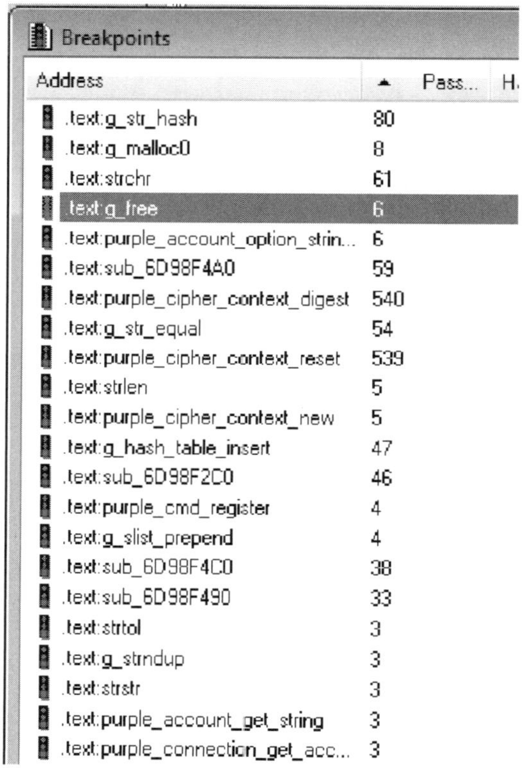

If we eliminate all the functions that didn't get called at least once, we're down to 133 functions. That cuts our search space down to 24% of the original. Many of the functions in the list are wrapper functions that simply call another function in the list. I like to start with the functions that are called often, but not hundreds of times. Just glancing at each of the functions, you should fairly quickly come to the function shown in Figure 7.25.

Figure 7.25 Frequently Called Functions

```
sub_6D9897C0 proc near

var_458= dword ptr -458h
var_454= dword ptr -454h
var_450= dword ptr -450h
var_44C= dword ptr -44Ch
var_448= dword ptr -448h
ReadBuffer= dword ptr -42Ch
var_41C= dword ptr -41Ch
var_418= dword ptr -418h
arg_0= dword ptr  8

push    ebp
mov     edx, 400h
mov     ebp, esp
push    edi
push    esi
lea     esi, [ebp+var_418]
push    ebx
sub     esp, 44Ch       ; void *
mov     eax, [ebp+arg_0]
mov     eax, [eax+1Ch]
mov     [ebp+var_41C], eax
mov     [esp+458h+var_450], edx
mov     [esp+458h+var_454], esi
mov     eax, [eax+4]
mov     [esp+458h+var_458], eax
call    wpurple_read
test    eax, eax
mov     ebx, eax
jl      loc_6D989B17
```

It's always good to have dumps of the protocol in front of you for comparison. This is the first message my client sent to the server in hex.

```
59 4D 53 47 00 0F 00 00 13 00 57 00 00 00 00 00 00 00 00 31 C0 80 69 64 61 70 6C
75 67 69 6E 31 32 33 34 35 C0 80
```

The call to *wpurple_read* makes this function a good candidate to start reverse engineering. The function copies the data into another buffer, makes sure it's longer than four bytes, and then drops into the main processing blocks. The first protocol processing block is shown in Figure 7.26.

Figure 7.26 First Protocol Processing Block

```
loc_6D989877:
cld
mov     esi, [ebp+ReadBuffer]
mov     eax, offset aYmsg ; "YMSG"
cmp     ecx, ecx
mov     edi, eax
repe cmpsb
jz      loc_6D989950
```

It compares the first four bytes of the read buffer to the string "YMSG" and exits if it doesn't match. The YMSG must be used as a magic number, as shown in Figure 7.27.

Figure 7.27 YMSG Magic Number

```
loc_6D989950:
mov     edx, [ebp+ReadBuffer]
mov     esi, offset aYahoo_0 ; "yahoo"
movzx   eax, byte ptr [edx+8]
movzx   edx, byte ptr [edx+9]
mov     [esp+458h+var_448], ebx
mov     [esp+458h+SrcBuffer], esi
shl     eax, 8
lea     edi, [eax+edx]
mov     [esp+458h+MessageLen], edi
mov     eax, offset aDBytesToReadRx ; "%d bytes to read, rxlen is %d\n"
mov     [esp+458h+BufferLen], eax
mov     [esp+458h+DestBuffer], 1
call    purple_debug
mov     ecx, [ebp+var_41C]
lea     eax, [edi+14h]
cmp     [ecx+0Ch], eax
jl      short loc_6D989945
```

The next check looks like a length field. It pulls bytes 8 and 9 out of the packet, shifts byte 8 to the left, and adds them together. So we can guess that the length field is a big-endian short. That leaves our structure something like:

```
struct login_packet{
       char Magic[4];           /*Always YMSG*/
       char unknown1;
       char unknown2;
       char unknown3;
       char unknown4;
       unsigned short Len;      /*Big Endian*/
}
```

The process and number of function calls involved is greater in the Yahoo! Instant Messenger protocol than in the example protocol used earlier in the chapter, but extracting the protocol from the binary is the same. Input comes from a read and then is processed through a set of function calls. With a little work, you should be able to reverse the rest of the protocol.

Chapter 8

Advanced Walkthrough

Solutions in this chapter:

- **Reversing Malware**

Introduction

So now you've read the book, and should be able to do some of this on your own. In this chapter, we're going to look at a real piece of hostile code. The hostile code we're going to use is *real*; this means that you should be especially careful when dealing with it yourself because you could possibly do serious harm to your computer and your network. Please be positive you are authorized to analyze the application in your environment. I highly suggest the use of some form of virtualization software, such as VMware. One thing I especially like about VMware is the ability to take snapshots, which allows me to get to specified points and make a restore point. This makes things like travel and close of the business day easier as you don't have to worry about someone unplugging your computer or needing to take your laptop home with you. All that said, be careful and bear in mind that you and only you are responsible for your actions. In that sense, when we say advanced walkthrough, we don't mean so much that the content is especially advanced, but rather that we're taking all of the pieces and putting them together to form a coherent picture of what you can expect to be doing as a reverse engineer.

NOTE

Just like anything else, reverse engineering is something you will get more comfortable as you gain more experience doing it. Thus, once the base knowledge is there, it's really just a matter of doing it to become good. It can be problematic trying to find employment in the field until you're more qualified to do it. That's where the beauty of the internet comes in. There are many organizations and Web sites that have challenges, for instance the Honey Net project is fairly well know for presenting incident response type challenges. This can be fun and is a good way to get your notoriety levels increased, however the focus is rarely on reverse engineering.

There are a couple of especially good websites. The first and most obvious is http://crackmes.de. This website is full of user-submitted programs with various objectives for reversing. For instance, they have unpackme's where the challenge is to unpack the program; crackme's which simulates cracking commercial software and so on. They have challenges for multiple operating systems and have various degrees of difficulty.

The second Web site is more of a forum, and there are some challenges but its one of the best places to find likeminded people, http://community.reverse-engineering.net/. This is a forum that has a plethora of information and will give you access to smart people.

The third is the Offensive Computing website http://www.offensive-computing.com/, this Web site among other things maintains a repository of known malware and serves as a good site for cross-referencing anything you might find, and a good place to brush up on your reversing malware skills using the real deal.

Finally, there is another Web site known as OpenRCE, http://www.openrce.org, it's run by a fairly accomplished reverse engineer from TippingPoint named Pedram Amini, and it contains forums and plugins, scripts and so on. There also is a searchable listing of call chains for the main system libraries, which comes in handy more often than you think.

Reversing Malware

As I said earlier, this program really isn't malware per se but rather ad-ware (it's at least somewhat safer that way), but it is a real program that you can expect to find in the wild, especially if you work on an incident response team or similar. Furthermore, many people would include usage of tools like regmon, filemon and so on, and these applications have valid use especially in the business world where it's always a race against time. However, that's not reverse engineering, we're engineers not fortune tellers and if given the time there is really no reason to have to do this, you know what the program is doing while it does it, so there should rarely be a technical reason to have to use these tools, unless you're not actually reversing the binary. The other reason for this is that doing things like that is borderline dangerous, you let the program have control and you really can't know what it did, there is nothing that says you need to touch the filesystem, or registry, or standard API. Ultimately, if given the right application these methods will fail, plus this is a reverse engineering book. That said, you should probably know how to use these tools as there will be any number of times where the time allocated by management just won't allow you to do the job correctly-time is money. Without further ado though, let's just right into the application and start looking at it. You can download this code from www.syngress.com/solutions. The password to open the file (courtesy of Dan Kaminsky) is '!DANGER!-INFECTEDMALWARE!DANGER!'.

Now as we examine the beginning of this code, it's pretty obvious that this is a non-standard entry, this is usually a sign of self-modifying code in some form or fashion, often its evidence of a packer. We have at least three tell-tale hints in this section. The first and most obvious is the name of the section we're in is UPX1, which is standard for a program packed with UPX. The next is the hexadecimal data preceding our entry point. This is pretty typical for both packed and encrypted code/strings/et cetera, and finally the last sign—and this one takes a little bit (but not a lot) of experience, is the pusha instruction. UPX is trivial to bypass as it saves all the registers with a pusha, then restores them with a popa followed by an unconditional jump. Thus it's just a matter of breaking at the popa/jmp and then continuing from there or dumping the data to disk. We really should confirm that what we expect is what's actually going on. We can do this by examining the code and/or single-stepping through it, and then confirming that the popa/jmp combination is actually present. You would normally do the first by examining control flow; dynamically if necessary. We're not going to do that here because there is really no point in including 15 pages of me single stepping through UPX unpacking an application to confirm that control doesn't jump off into someplace weird. I wouldn't recommend doing that in general, and overall I would highly advise doing analysis in VMware or something similar.

Figure 8.1 popa/jmp Instruction

```
UPX1:0040A06A loc_40A06A:                          ; CODE XREF: UPX1:0040A062↑j
UPX1:0040A06A                 mov     ecx, 0AEF24857h
UPX1:0040A06F                 push    ebp
UPX1:0040A070                 call    dword ptr [esi+0A484h]
UPX1:0040A076                 or      eax, eax
UPX1:0040A078                 jz      short loc_40A081
UPX1:0040A07A                 mov     [ebx], eax
UPX1:0040A07C                 add     ebx, 4
UPX1:0040A07F                 jmp     short loc_40A059
UPX1:0040A081 ; ---------------------------------------------------------------
UPX1:0040A081
UPX1:0040A081 loc_40A081:                          ; CODE XREF: UPX1:0040A078↑j
UPX1:0040A081                 call    dword ptr [esi+0A488h]
UPX1:0040A087
UPX1:0040A087 loc_40A087:                          ; CODE XREF: UPX1:0040A040↑j
UPX1:0040A087                 popa
UPX1:0040A088                 jmp     near ptr word_40395E
UPX1:0040A088 ; ---------------------------------------------------------------
```

As we can see in Figure 8.1, what we expected is there exactly- at loc_40A081. We have the popa/jmp instruction. Typically, in order to use IDA's unpacker plug-in we would need the Original Entry Point (OEP). However, because UPX wasn't exactly made to be obfuscated/used for malicious software, the OEP is obvious and is word_40395E, let's just take a quick look at that address and just make sure this is an address that makes sense. (See Figure 8.2).

Figure 8.2 Examining the Address

```
UPX1:00408000          dd 0AF4BF3Ch, 59BF2E05h, 0F72CB9FBh, 37400189h, 0FBFF5060h
UPX1:00408000          dd 15E16F1Bh, 5953426Ch, 0C98309A5h, 6028BFFFh, 0B7BF2B0Eh
UPX1:00408000          dd 0F2C0337Fh, 2BD1F7AEh, 0F78BD1FEh, 11635813h, 7FDBF7CCh
UPX1:00408000          dd 0C14FCA8Bh, 0A5F302E9h, 3E18307h, 242CA4F3h, 0C1B58B99h
UPX1:00408000          dd 272CF72Ah, 0B2AF2A16h, 20BF4F60h, 582A1860h, 7746B363h
UPX1:00408000          dd 7C2CC120h, 3E9B2141h, 5FC8918Ch, 1CEC8111h, 0F68B530Eh
UPX1:00408000          dd 9CEEFB5Bh, 55072824h, 40242D8Bh, 0DB85C43Ah, 17C3840Fh
UPX1:00408000          dd 0FCDED982h, 24B48B00h, 53036A30h, 0A268E456h, 0C2B67B9Bh
UPX1:00408000          dd 8E1B0D0Ch, 8DE562FBh, 0E1980D82h, 0C2182494h, 9212FA6Bh
UPX1:00408000          dd 70C16BAEh, 0C0507222h, 590FB970h, 0BDC2FA1Bh, 1C24848Dh
UPX1:00408000          dd 0BBCCE038h, 8C8B2F9Fh, 51933824h, 532015FFh, 0FDAC10BEh
UPX1:00408000          dd 20868EEh, 0FFE92604h, 92B48DD5h, 16C7D9DDh, 8C8D20EBh
UPX1:00408000          dd 5118041Fh, 16CD7756h, 45C207ECh, 80B73D1Ch, 0B35B2A73h
UPX1:00408000          dd 6468DDB9h, 458688Ah, 6501816h, 6685F058h, 0D005E62Bh
UPX1:00408000          dd 815B5D1Bh, 9B3982C4h, 0CC216FDh, 5256CA00h, 0D8397718h
UPX1:00408000          dd 7E1274EEh, 50130528h, 7C10EB53h, 3E3B3305h, 2C55111h
UPX1:00408000          dd 35CE355Ch, 0A7289B9Fh, 7150029Ch, 63962CCFh, 10485884h
UPX1:00408000          dd 0E5D0EAEh, 8316CD87h, 254CBF38h, 9EE0DC58h, 682B1EF0h
```

Looking at this address, it points into an offset into the long sequence of hexadecimal characters, which is exactly what we expect as the entry point doesn't typically point to the very beginning of the original file. So having a reasonable assurance that what we're looking at actually is UPX, let's see if we can get it unpacked. All recent versions of IDA (since 4.8) include a plug-in that is supposed to perform universal unpacking. We're going to briefly walk through that just in case you're not familiar with it. However, just like the previous chapter we're not going to get into unpacking too much.

Figure 8.3 Universal PE Unpacker

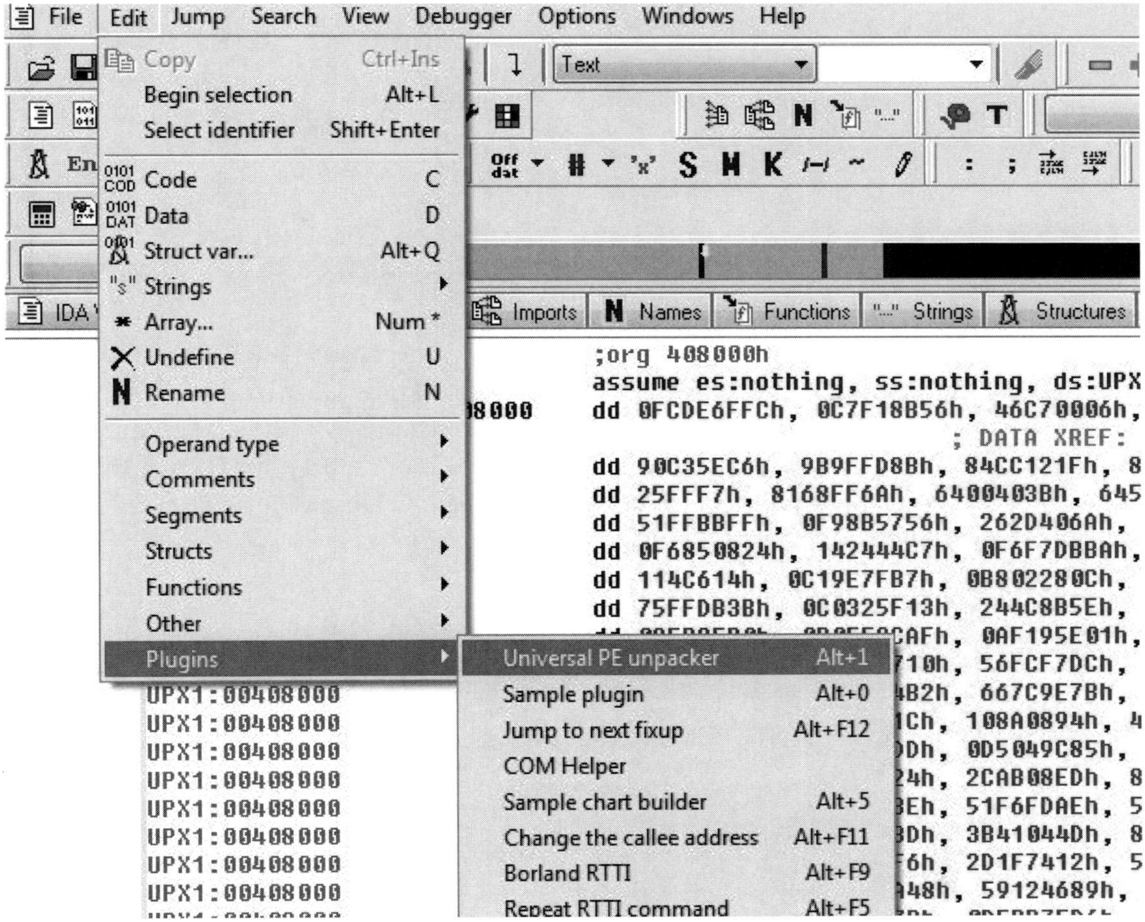

From inside of IDA, getting to the plug-in is pretty simple, it can be found under **Edit > Plugins > Universal PE unpacker** or you can simply hit Alt+1 (see Figure 8.3). From there, a warning letting you know that if you're not careful you could own yourself is displayed, and the dialog seen in Figure 8.4 appears.

Figure 8.4 Uunp Parameters

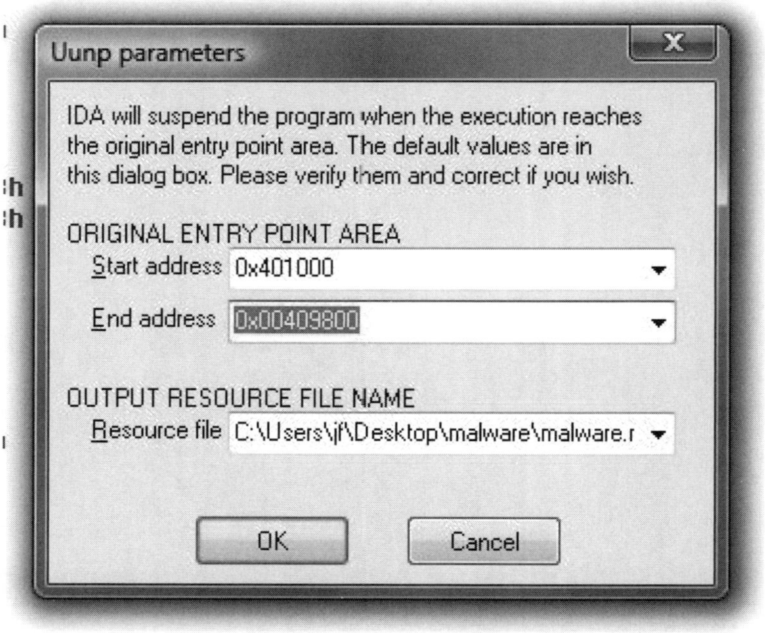

As you can see, knowing the exact OEP is not entirely necessary. Basically, this technology is not very advanced in the sense that the debugger doesn't know what it's doing. All it does is watch for execution in the range including/between Start Address and End Address. In this case, I modified the default End address parameter to reflect our binary and then clicked **OK**. When IDA detects that the necessary conditions have been met, it prompts you with the dialog shown in Figure 8.5.

Figure 8.5 Confirm Dialog

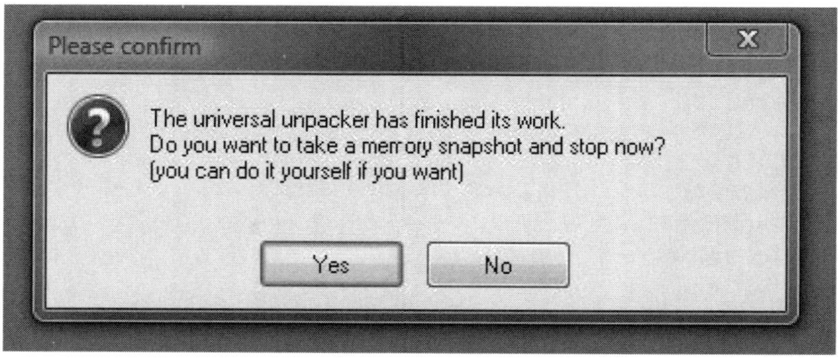

Once **OK** is clicked, IDA terminates the debugger and does a lot of stuff in the background for you. For example it rebuilds the imports, reanalyzes code flow, et cetera. Once all of that is complete we're looking at the unpacked version of the code, which is a bit closer to what we expect to see (Figure 8.6).

Figure 8.6 Unpacked Version of Code

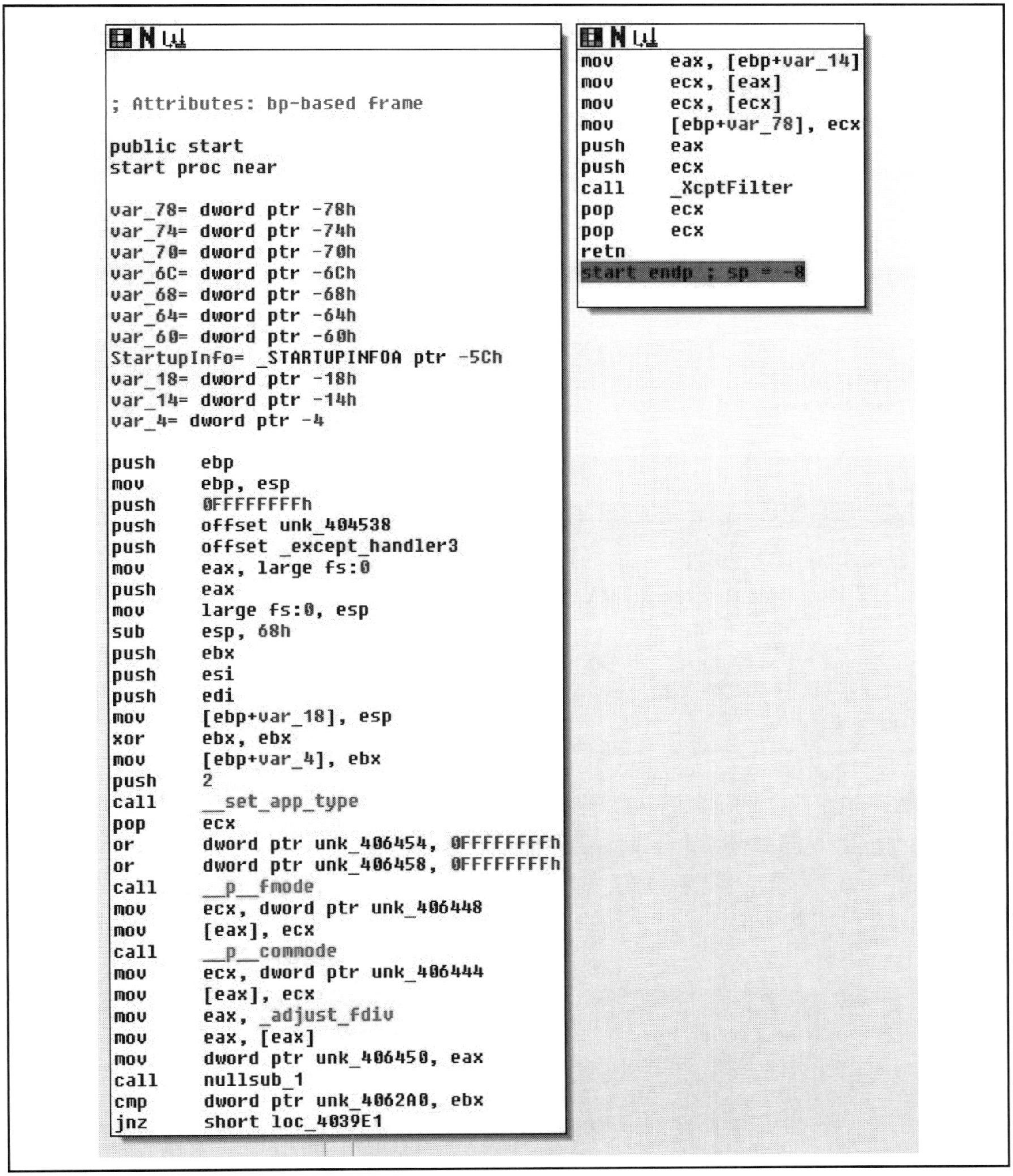

Examining the code, we see a typical routine entry, however if you note references to routines like nullsub_1, and if you were to dig a bit deeper you'd find references like those shown in Figure 8.7.

Figure 8.7 References to Routines

```
.idata:00404174    extrn __imp__941:dword  ; DATA XREF: _941↑r
.idata:00404178    extrn __imp__1576:dword ; DATA XREF: _1576↑r
.idata:0040417C    extrn __imp__2976:dword ; DATA XREF: _2976↑r
.idata:00404180    extrn __imp__3081:dword ; DATA XREF: _3081↑r
.idata:00404184    extrn __imp__2985:dword ; DATA XREF: _2985↑r
.idata:00404188    extrn __imp__3262:dword ; DATA XREF: _3262↑r
.idata:0040418C    extrn __imp__3136:dword ; DATA XREF: _3136↑r
.idata:00404190    extrn __imp__1776:dword ; DATA XREF: _1776↑r
.idata:00404194    extrn __imp__101_1:dword ; DATA XREF: _101_1↑r
.idata:00404198    extrn __imp__101_2:dword ; DATA XREF: _101_2↑r
.idata:0040419C    extrn __imp__101_3:dword ; DATA XREF: _101_3↑r
.idata:004041A0    extrn __imp__5714:dword ; DATA XREF: _5714↑r
.idata:004041A4    extrn __imp__5289:dword ; DATA XREF: _5289↑r
.idata:004041A8    extrn __imp__5307:dword ; DATA XREF: _5307↑r
.idata:004041AC    extrn __imp__4698:dword ; DATA XREF: _4698↑r
.idata:004041B0    extrn __imp__4079:dword ; DATA XREF: _4079↑r
.idata:004041B4    extrn __imp__2725:dword ; DATA XREF: _2725↑r
.idata:004041B8    extrn __imp__5302:dword ; DATA XREF: _5302↑r
.idata:004041BC    extrn __imp__5300:dword ; DATA XREF: _5300↑r
.idata:004041C0    extrn __imp__3346:dword ; DATA XREF: _3346↑r
.idata:004041C4    extrn __imp__2396:dword ; DATA XREF: _2396↑r
.idata:004041C8    extrn __imp__5199:dword ; DATA XREF: _5199↑r
.idata:004041CC    extrn __imp__1089:dword ; DATA XREF: _1089↑r
.idata:004041D0    extrn __imp__3922:dword ; DATA XREF: _3922↑r
.idata:004041D4    extrn __imp__5731:dword ; DATA XREF: _5731↑r
.idata:004041D8    extrn __imp__2512:dword ; DATA XREF: _2512↑r
.idata:004041DC    extrn __imp__2554:dword ; DATA XREF: _2554↑r
.idata:004041E0    extrn __imp__4486:dword ; DATA XREF: _4486↑r
.idata:004041E4    extrn __imp__6375:dword ; DATA XREF: _6375↑r
```

As you can see, IDA has not reanalyzed all of the imported functions. This happened because the FLIRT signatures that IDA typically runs over the code were not applied. This is not a huge issue because we can do it ourselves without much effort. If you're unfamiliar with this, that's okay, as we're going to talk about it some.

Figure 8.8 Load File

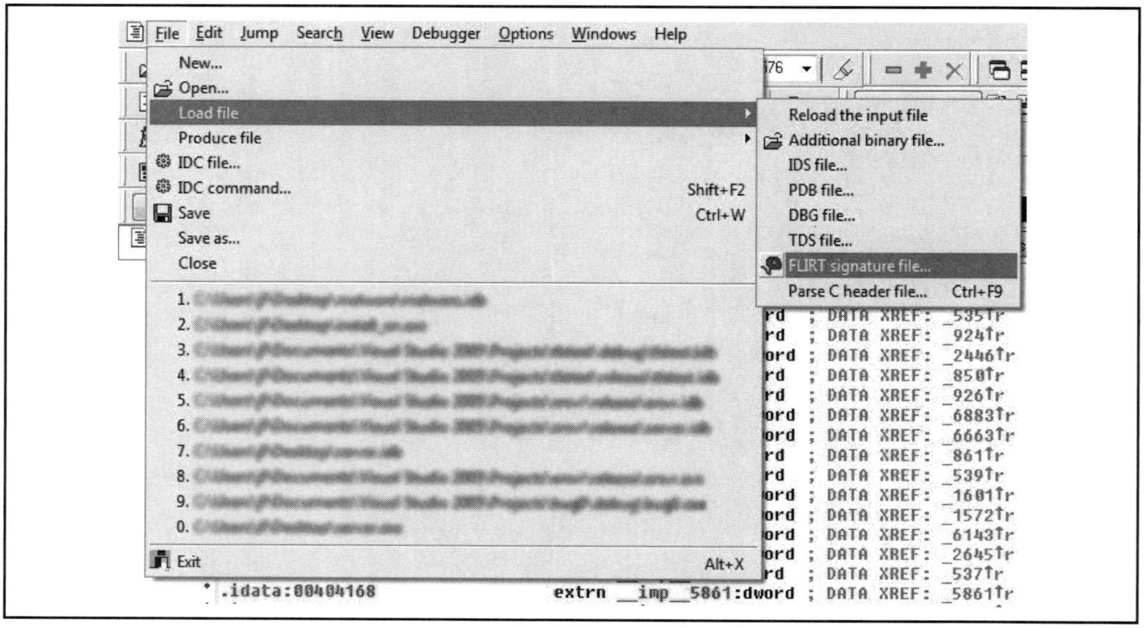

In order to get to the options for loading FLIRT signature files, you go to **File > Load File > FLIRT signature file** (Figure 8.8), from this menu you get the prompt in Figure 8.9.

Figure 8.9 List of Available Library Modules

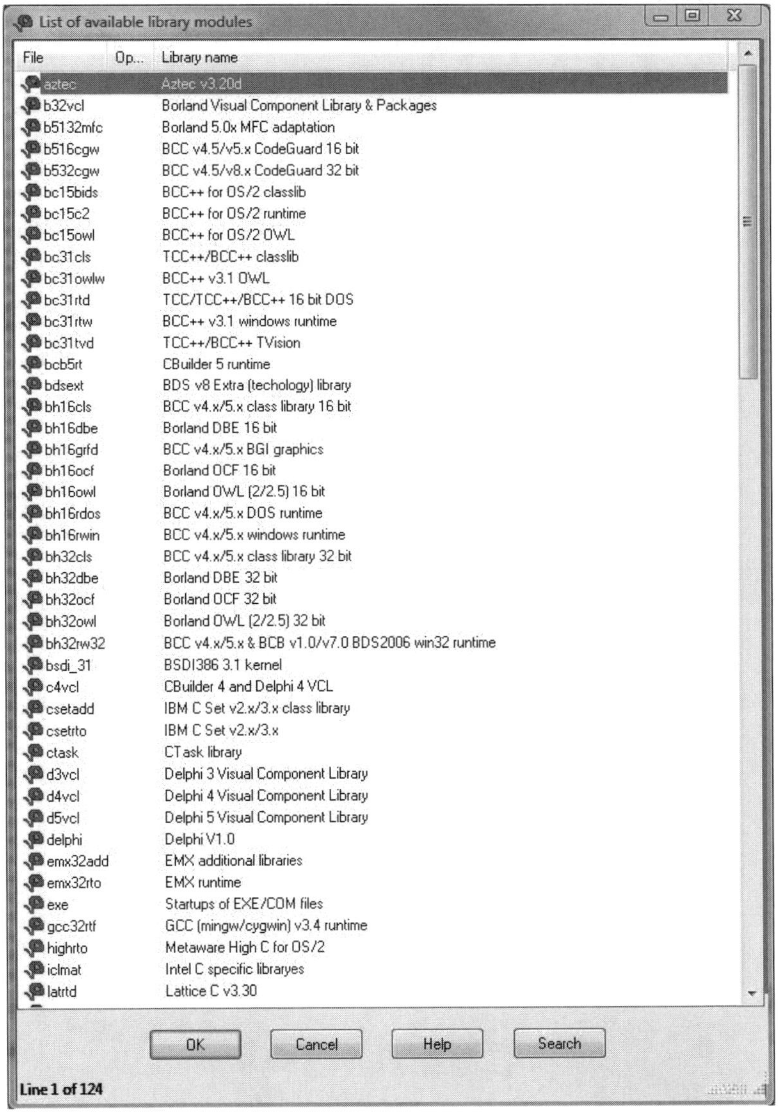

Looking at the prompt, we get some idea of what it does; specifically we see a list of compilers/libraries. Thus the FLIRT signature file applies characteristics known to exist for a given compiler. But how do we know what compiler is in use? This at first seems like it would be implausibly hard, but generally it's pretty easy and failing that it's where some experience comes in. Usually, we can look through the strings in a given binary and the compiler inserted a string advertising its use. Failing that we can look at the generated code and try to guess. What to look for exactly when examining the code is a bit beyond the scope of this book. You'll get a better idea of how to do that as time progresses. For instance however, if you've looked at much code generated by GCC you will probably note that it likes to allocate more space on the stack than necessary, then in the next instruction or two it will correct that, you're basically looking for things like that. In our instance, these are our strings shown in Figure 8.10.

Figure 8.10 Strings

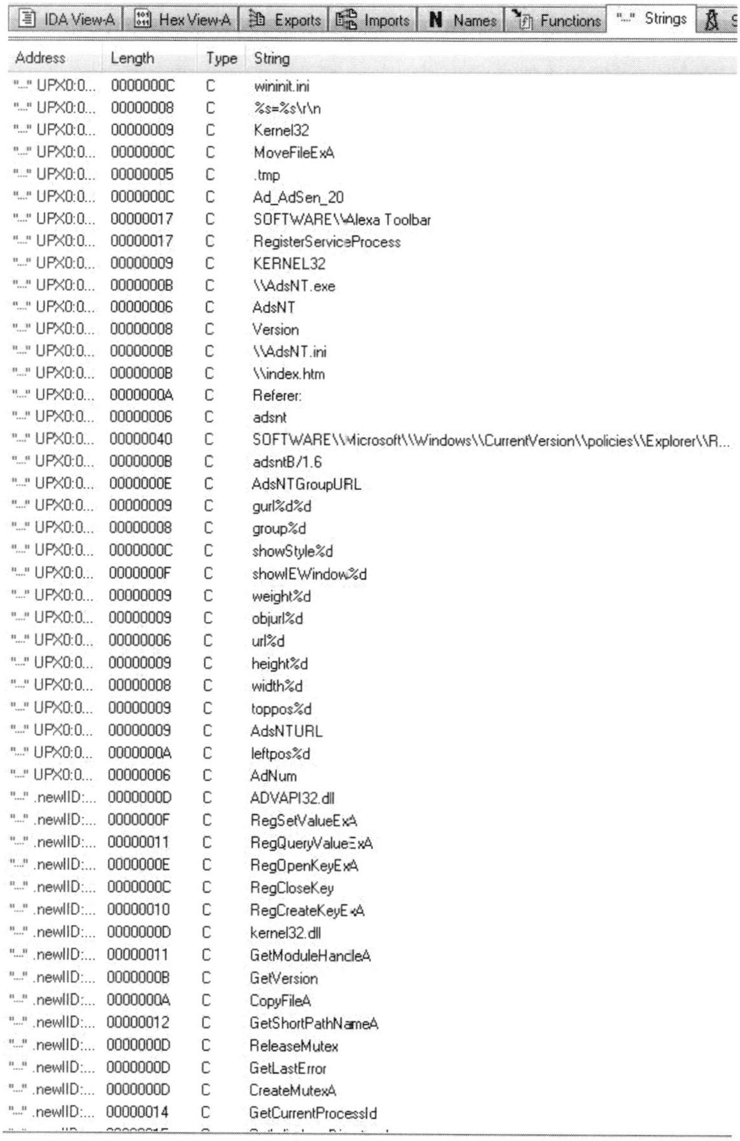

IDA View-A | Hex View-A | Exports | Imports | Names | Functions | "..." Strings

Address	Length	Type	String
"..." UPX0:0...	0000000C	C	wininit.ini
"..." UPX0:0...	00000008	C	%s=%s\r\n
"..." UPX0:0...	00000009	C	Kernel32
"..." UPX0:0...	0000000C	C	MoveFileExA
"..." UPX0:0...	00000005	C	.tmp
"..." UPX0:0...	0000000C	C	Ad_AdSen_20
"..." UPX0:0...	00000017	C	SOFTWARE\\Alexa Toolbar
"..." UPX0:0...	00000017	C	RegisterServiceProcess
"..." UPX0:0...	00000009	C	KERNEL32
"..." UPX0:0...	0000000B	C	\\AdsNT.exe
"..." UPX0:0...	00000006	C	AdsNT
"..." UPX0:0...	00000008	C	Version
"..." UPX0:0...	0000000B	C	\\AdsNT.ini
"..." UPX0:0...	0000000B	C	\\index.htm
"..." UPX0:0...	0000000A	C	Referer:
"..." UPX0:0...	00000006	C	adsnt
"..." UPX0:0...	00000040	C	SOFTWARE\\Microsoft\\Windows\\CurrentVersion\\policies\\Explorer\\R...
"..." UPX0:0...	0000000B	C	adsntB/1.6
"..." UPX0:0...	0000000E	C	AdsNTGroupURL
"..." UPX0:0...	00000009	C	gurl%d%d
"..." UPX0:0...	00000008	C	group%d
"..." UPX0:0...	0000000C	C	showStyle%d
"..." UPX0:0...	0000000F	C	showIEWindow%d
"..." UPX0:0...	00000009	C	weight%d
"..." UPX0:0...	00000009	C	objurl%d
"..." UPX0:0...	00000006	C	url%d
"..." UPX0:0...	00000009	C	height%d
"..." UPX0:0...	00000008	C	width%d
"..." UPX0:0...	00000009	C	toppos%d
"..." UPX0:0...	00000009	C	AdsNTURL
"..." UPX0:0...	0000000A	C	leftpos%d
"..." UPX0:0...	00000006	C	AdNum
"..." .newIID:...	0000000D	C	ADVAPI32.dll
"..." .newIID:...	0000000F	C	RegSetValueExA
"..." .newIID:...	00000011	C	RegQueryValueExA
"..." .newIID:...	0000000E	C	RegOpenKeyExA
"..." .newIID:...	0000000C	C	RegCloseKey
"..." .newIID:...	00000010	C	RegCreateKeyExA
"..." .newIID:...	0000000D	C	kernel32.dll
"..." .newIID:...	00000011	C	GetModuleHandleA
"..." .newIID:...	0000000B	C	GetVersion
"..." .newIID:...	0000000A	C	CopyFileA
"..." .newIID:...	00000012	C	GetShortPathNameA
"..." .newIID:...	0000000D	C	ReleaseMutex
"..." .newIID:...	0000000D	C	GetLastError
"..." .newIID:...	0000000D	C	CreateMutexA
"..." .newIID:...	00000014	C	GetCurrentProcessId

Unfortunately, we don't get to see any compiler strings so we have to take an educated guess. If worse comes to worse we can just start loading FLIRT signatures until we've resolved most everything. In our instance, this is mostly what happened. There are numerous signatures that matched against. The first and most common is the MS Visual C++ runtime signature, but then using MFC starter signatures turned up the most results. In the end, I managed to get most everything resolved, this particular binary was a little wonky, but that's entirely because it's using MFC 4.2. With all of that resolved, towards the end of the entry points function, we see that the call at the end has been resolved to a WinMain() routine, and when we look at that we see Figure 8.11.

Figure 8.11 WinMain() Routine

```
; Attributes: library function

; int __stdcall WinMain(HINSTANCE hInstance,HINSTANCE hPrevInstance,LPSTR lpCmdLine,int nShowCmd)
_WinMain@16 proc near

hInstance= dword ptr  4
hPrevInstance= dword ptr  8
lpCmdLine= dword ptr  0Ch
nShowCmd= dword ptr  10h

push    [esp+nShowCmd]
push    [esp+4+lpCmdLine]
push    [esp+8+hPrevInstance]
push    [esp+0Ch+hInstance]
call    ?AfxWinMain@@YGHPAUHINSTANCE__@@0PADH@Z ; AfxWinMain(HINSTANCE__ *,HINSTANCE__ *,char *,int)
retn    10h
_WinMain@16 endp
```

The one interesting thing is that the code is using MFC. I believe out of all the malware I've analyzed, this is only the second or third that employs MFC. Now, I've managed to spend a fairly significant portion of my life without learning much of anything about MFC, and if we're lucky we can continue that trend today. So I set a breakpoint on the call and single stepped into the call, hoping that I wouldn't hit a ton of MFC kludge (See Figure 8.12).

Figure 8.12 General Registers

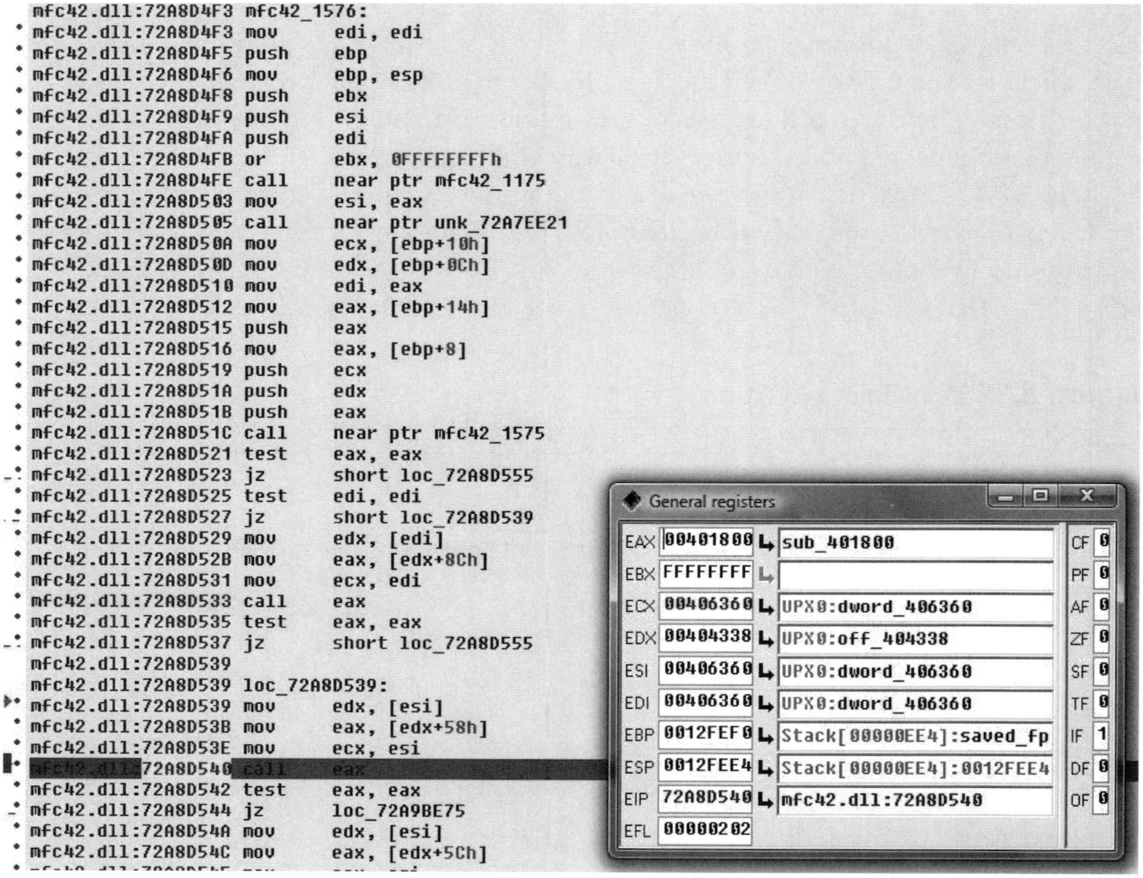

This was a fairly simple function and getting to what we were looking for was rather painless. Because this is malware I've never analyzed and I'm not entirely positive what it does, nor the mechanics of AfxWinMain(), I stepped over the internal calls to other MFC APIs. I had noticed the two calls to the EAX register and figured that one of them was probably what I was looking for. The second one was, as we've noted it hands control back into the application to sub_401800. Now we're going to terminate the debugger and continue static analysis at sub_401800. After terminating the debugger, I simply went to the **functions** tab, which can be found in between the **strings** and **names** tabs—if it's not there you may have closed it and you can reopen it by going to **View > Open Subviews > Functions**, or hitting **Shift+F3**. Once there we see the code in Figure 8.13.

Figure 8.13

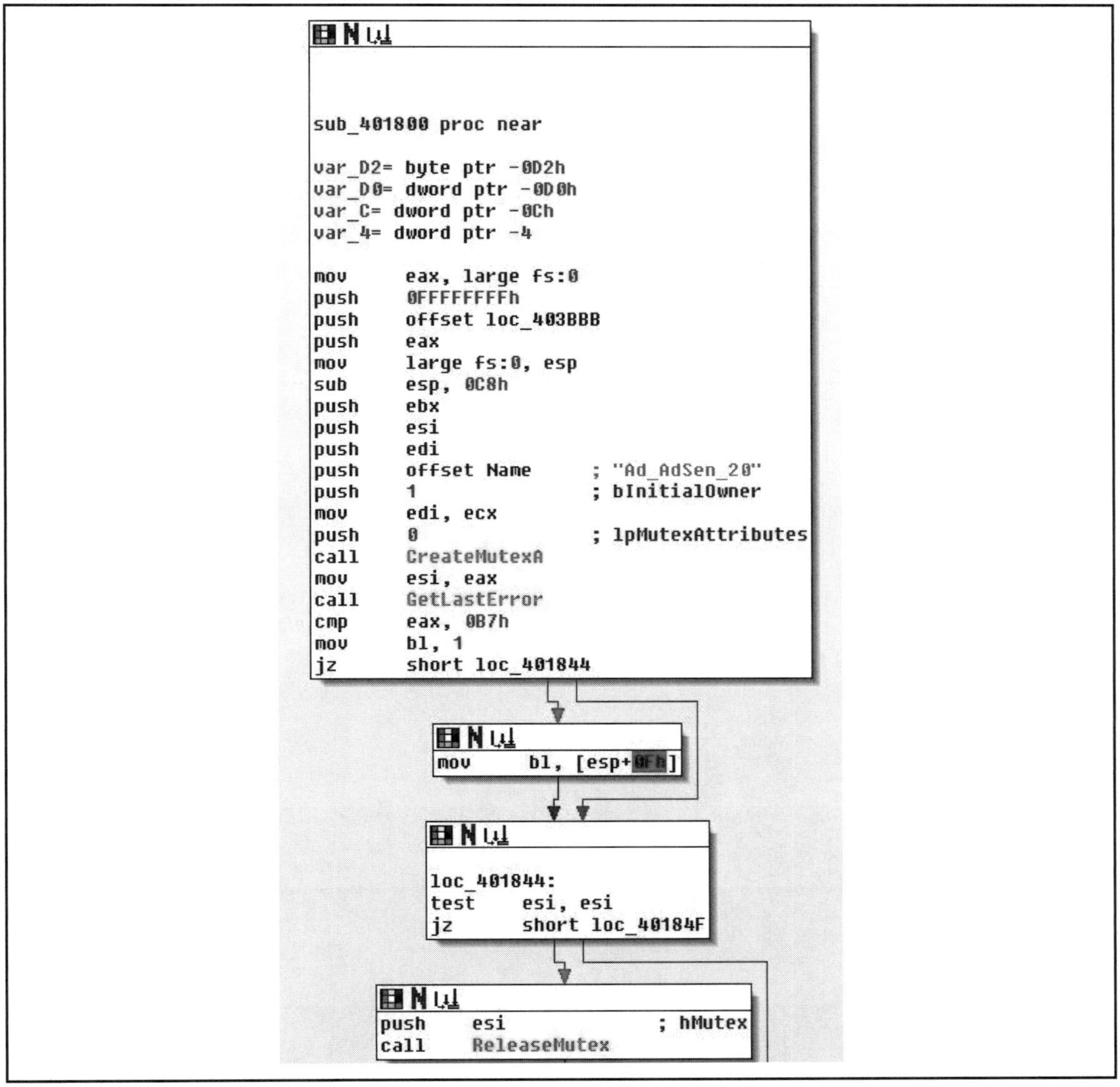

As we can see, the subroutine takes no arguments, and has relatively few arguments. The prologue of the procedure is pretty standard. The first real code we see is a call to CreateMutexA() with an argument of "Ad_AdSen_20". This is potentially used to synchronize access between threads, and we will know for sure as we continue into the application. However, it's quite likely that this mutex actually prevents double-infection and from multiple copies of the application from trampling itself. Another aspect is that if we Googled for that string, we'd probably find what information we were looking for and we basically just acted as a human anti-virus application. However, we will just assume that we didn't get any results and continue analyzing.

What we see here after the call to CreateMutexA() is that the return value is copied into the ESI register and GetLastError() is called. This return value is compared against the constant value 0xB7, or 183, which is the value for ERROR_ALREADY_EXISTS. This is exactly what we're expecting to see if they're trying to prevent themselves from trampling themselves. From there, the value of 1 is copied into the lower 8-bits of the EBX register, and then we branch depending on the return value. If the condition was true, that CreateMutexA() failed, then we branch to loc_401844, otherwise we copy a value from the stack into the low-order 8-bits of EBX. From there, we test to see if the ESI register is non- zero (CreateMutex() returns NULL on error). If this is the case we branch off to loc_40184F, otherwise we release ownership of the mutex. Moving on down the function we find the code in Figure 8.14.

Figure 8.14

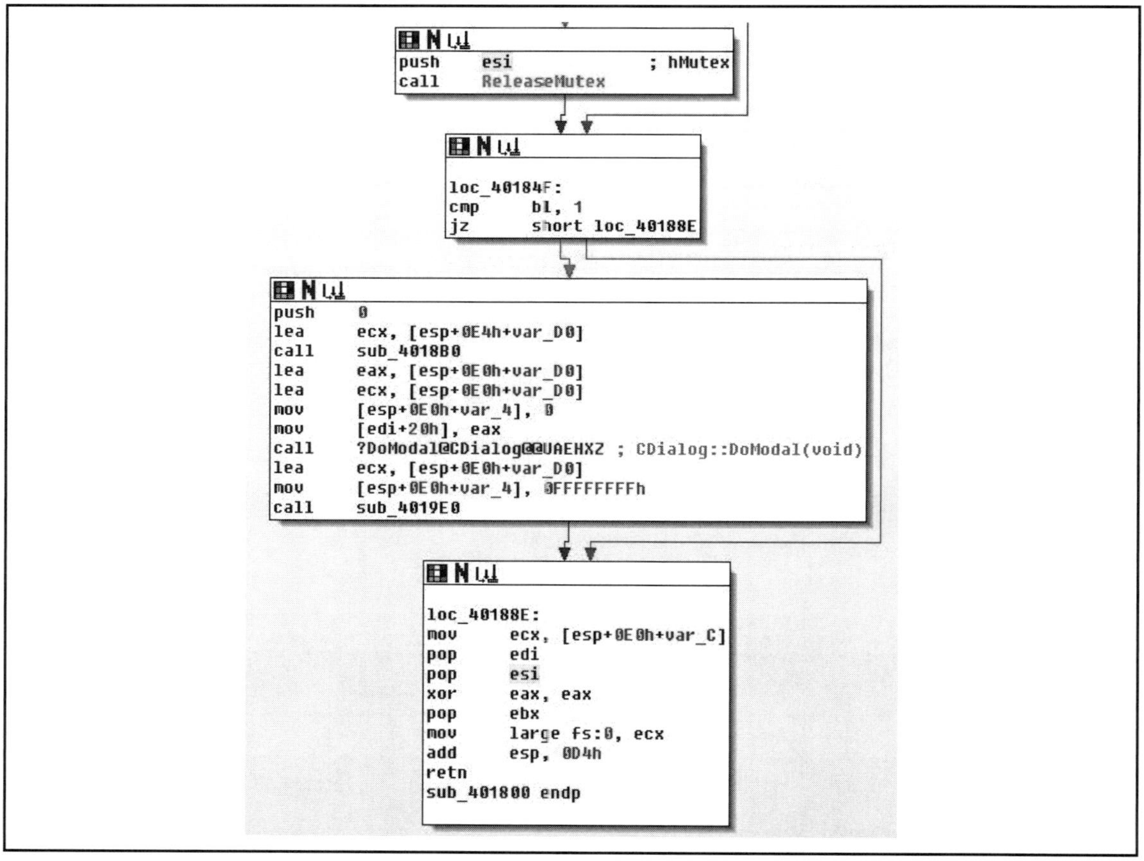

Moving on through the code, we see the potential call to ReleaseMutex(). Following that, at loc_40184F the low-order 8-bits of the EBX register is checked against the value of 1. If you remember correctly this constant value was moved in and remained in there if the call to CreateMutexA() failed, thus the code at loc_40184F simply checks to see if the mutex was successfully created/ownership was giving to the calling thread. If the call failed, execution control is handed off to loc_40188E, otherwise it continues down. Assuming there was no failure, then zero is pushed onto the stack as a parameter to sub_4018B0(). We also note that the address of an offset in var_D0 is placed in the ECX register immediately before calling the routine, and then before it's used again the value is overwritten with a different offset. There are two possible explanations here, the first is that a fastcall calling convention is being employed, and the second is that we are calling a method inside a class. Because the first argument is pushed on the stack, instead of being placed in the ECX register (with the second being in the EDX register), we can assume that we are calling an instance method and that what exists in the ECX register is the *this* pointer. We'll now take a look at sub_4018B0() and see what it does and try to confirm our suspicions about the ECX register shown in Figure 8.15.

Figure 8.15 ECX Register

```
sub_4018B0:
push    0FFFFFFFFh
push    offset loc_403C3C
mov     eax, large fs:0
push    eax
mov     large fs:0, esp
push    ecx
mov     eax, [esp+10h+arg_0]
push    esi
push    edi
mov     esi, ecx
push    eax
push    66h
mov     [esp+20h+var_10], esi
call    ??0CDialog@@QAE@IPAVCWnd@@@Z; CDialog::CDialog(uint,CWnd *)
lea     ecx, [esi+60h]
mov     [esp+18h+var_4], 0
call    ??0CString@@QAE@XZ; CString::CString(void)
lea     ecx, [esi+64h]
mov     byte ptr [esp+18h+var_4], 1
call    ??0CString@@QAE@XZ; CString::CString(void)
lea     ecx, [esi+68h]
mov     byte ptr [esp+18h+var_4], 2
call    sub_401000
lea     ecx, [esi+78h]
mov     byte ptr [esp+18h+var_4], 3
call    ??0CStringArray@@QAE@XZ; CStringArray::CStringArray(void)
lea     ecx, [esi+90h]
mov     byte ptr [esp+18h+var_4], 4
call    ??0CStringArray@@QAE@XZ; CStringArray::CStringArray(void)
lea     ecx, [esi+0A4h]
mov     byte ptr [esp+18h+var_4], 5
call    ??0CString@@QAE@XZ; CString::CString(void)
lea     ecx, [esi+0A8h]
mov     byte ptr [esp+18h+var_4], 6
call    ??0CStringArray@@QAE@XZ; CStringArray::CStringArray(void)
lea     edi, [esi+0BCh]
mov     byte ptr [esp+18h+var_4], 7
mov     ecx, edi
call    ??0CString@@QAE@XZ; CString::CString(void)
push    offset Default
mov     ecx, edi
mov     byte ptr [esp+1Ch+var_4], 8
mov     dword ptr [esi], offset off_404460
mov     dword ptr [esi+8Ch], 0
call    ??4CString@@QAEABV0@PBD@Z; CString::operator=(char const *)
mov     ecx, esi
call    sub_4035A0
push    0               ; pvReserved
call    CoInitialize
mov     ecx, [esp+18h+var_C]
mov     eax, esi
pop     edi
pop     esi
mov     large fs:0, ecx
add     esp, 10h
retn    4
```

As we enter sub_4018B0(), we see a fairly standard entry, and we note that this is employing the fastcall calling convention. After the procedure prologue, this also includes the setup of the SEH record. After this we see that the ECX register is saved and the parameter that was pushed onto the stack segment is copied into the EAX register. A copy of the ECX register is saved in the ESI register and the *this* pointer is saved in var_10 with finally then the constructor for the CDialog class being called. The two arguments passed to the constructor indicate that we don't want a particular parent window, or rather that the parent window should be that of the main application and the integer specifying a dialog template to use. Here we're just getting ready to setup a dialog box, which for being malware is a little odd to be honest. We'll keep looking through it though and eventually the pieces will come together. Moving on through the routine, we see an offset from the current 'this' pointer is calculated and saved in the ECX register, then var_14 is initialized to zero and the constructor for CString::CString(void) is called. If you were to single step into that constructor you would see Figure 8.16, which helps make sense of what's going on in the binary.

Figure 8.16 Single Step Constructor

So knowing that in our context, the ECX register is an offset from the *this* pointer of the parent method. We can clearly see that a copy of it is made in the ESI register followed by a call to a near offset. However, interestingly enough, we see the apparent return value overwritten with the pointer that was passed into the method by the caller. Knowing that this is a constructor, we can almost look at this as being a placement new sort of thing. More likely however, is that our caller is a constructor in itself, which makes some sense as if you follow the CDialog::CDialog() call in the caller's frame. You will find a recursive call into sub_4018B0(). Another interesting aspect is that if you note, there are obviously several instance variables being instantiated in this method, to be more specific four instances of Cstring and three instances of CStringArray. However, the instance of CDialog is a little odd in that the 'this' pointer is not modified upon entry. This would at least imply that our class is some abstraction/interface/et cetera of the CDialog class.

Throughout the routine, we see some oddities. For instance the *this* pointer is saved into var_10, but the variable is never referenced again, var_4 is incremented after each new object is constructed, but never referenced again and so on. The point being that hopefully you're starting to get the picture about the different types of clues compilers leave laying around for you. Following this, we see the modification of the data at the *this* pointer—it being modified to off_404460. What is interesting is that the pointer to the last CString object is overwritten with a NULL pointer, leaving an object with no reference. Although, we will reserve judgment until the destructor is inspected to make that determination. The call to the assignment operator for the CString object is also simple to figure out, if you were to single step in, you'd see that quite simply this just ensures nil-termination for the string. Next, we see that the *this* pointer copy in the ESI register is copied into the ECX register and sub_4035A0() is called, implying it is another member method of this class.

Stepping into sub_4035A0() we see the code in Figure 8.17, and it becomes obvious that what we're viewing is a string initialization routine, and furthermore that a somewhat vague attempt at making it obfuscated has been made, or at least that appears to be the case. This presumption is based off of the number of calls to the operator+=and all of the arguments being string constants.

Figure 8.17

```
sub_4035A0 proc near
push    esi
push    edi
mov     edi, ecx
call    sub_403630
lea     esi, [edi+60h]
push    offset dword_40626C
mov     ecx, esi
call    ??YCString@@QAEABV0@PBD@Z ; CString::operator+=(char const *)
push    offset dword_406264
mov     ecx, esi
call    ??YCString@@QAEABV0@PBD@Z ; CString::operator+=(char const *)
push    offset dword_406260
mov     ecx, esi
call    ??YCString@@QAEABV0@PBD@Z ; CString::operator+=(char const *)
push    offset dword_406258
mov     ecx, esi
call    ??YCString@@QAEABV0@PBD@Z ; CString::operator+=(char const *)
lea     esi, [edi+64h]
push    offset dword_406254
mov     ecx, esi
call    ??YCString@@QAEABV0@PBD@Z ; CString::operator+=(char const *)
push    offset dword_406250
mov     ecx, esi
call    ??YCString@@QAEABV0@PBD@Z ; CString::operator+=(char const *)
push    offset dword_406244
mov     ecx, esi
call    ??YCString@@QAEABV0@PBD@Z ; CString::operator+=(char const *)
push    offset dword_40623C
mov     ecx, esi
call    ??YCString@@QAEABV0@PBD@Z ; CString::operator+=(char const *)
push    offset dword_406238
mov     ecx, esi
call    ??YCString@@QAEABV0@PBD@Z ; CString::operator+=(char const *)
push    offset dword_406230
mov     ecx, esi
call    ??YCString@@QAEABV0@PBD@Z ; CString::operator+=(char const *)
pop     edi
pop     esi
retn
sub_4035A0 endp
```

Upon entry into sub_3035A0(), we see a frameless method. That is to say that we do not see a strict procedure prologue or epilogue and we almost immediately call into sub_403630(), examining that yields a similar method as the one in Figure 8.18.

Figure 8.18

```
sub_403630 proc near
push    esi
push    edi
mov     edi, ecx
push    offset dword_40628C
lea     esi, [edi+60h]
mov     ecx, esi
call    CString__equals
push    offset dword_406284
mov     ecx, esi
call    ??YCString@@QAEABV0@PBD@Z ; CString::operator+=(char const *)
push    offset dword_406280
mov     ecx, esi
call    ??YCString@@QAEABV0@PBD@Z ; CString::operator+=(char const *)
push    offset dword_40627C
mov     ecx, esi
call    ??YCString@@QAEABV0@PBD@Z ; CString::operator+=(char const *)
push    offset dword_406278
mov     ecx, esi
call    ??YCString@@QAEABV0@PBD@Z ; CString::operator+=(char const *)
push    offset dword_406274
mov     ecx, esi
call    ??YCString@@QAEABV0@PBD@Z ; CString::operator+=(char const *)
lea     esi, [edi+64h]
push    offset dword_40628C
mov     ecx, esi
call    CString__equals
push    offset byte_406270
mov     ecx, esi
call    ??YCString@@QAEABV0@PBD@Z ; CString::operator+=(char const *)
pop     edi
pop     esi
retn
sub_403630 endp
```

If you are really interested, you can single step through these two methods and analyze them, however all they do is result in two CStrings with the values of "http://www.alxup.com/adsnt/AdsNT.ini" and "http://www.alxup.com/adsnt/AdsNT.exe" respectively.

Finally, upon returning back to the constructor method, we see a call to CoInitialize() and the *this* pointer is returned to the caller in the EAX register. Knowing what we know, we'll call this method we just analyzed derrivedDialog::derrivedDialog(). Moving back to the calling routine, the WinMain() function we have the code in Figure 8.19.

Figure 8.19

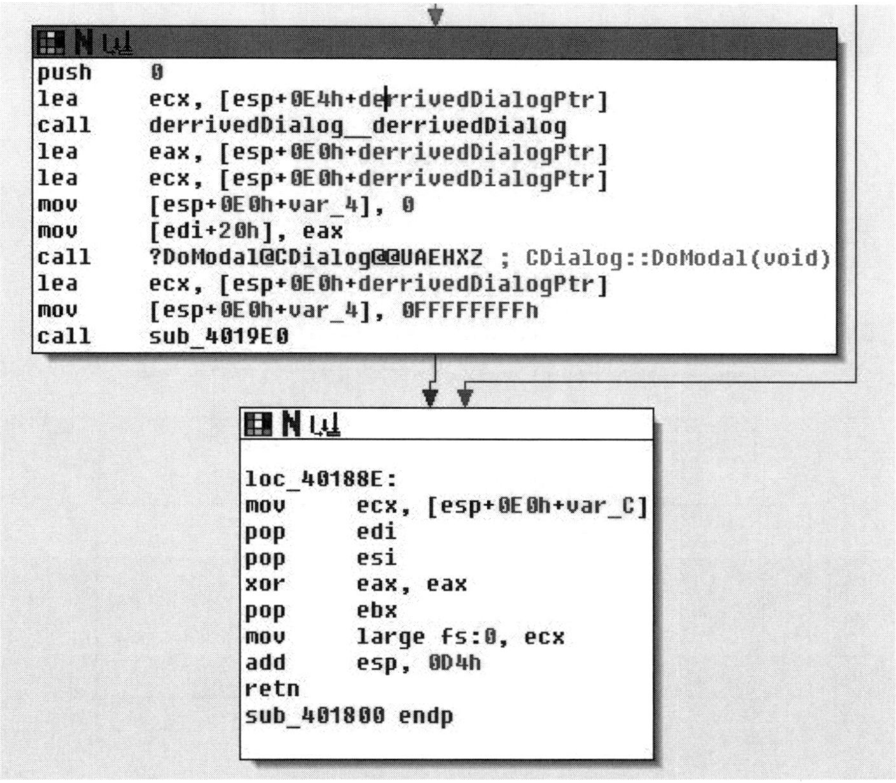

After the object construction, which is what we just analyzed we see that a copy of the pointer to the object is made in the EAX and ECX registers, that var_4 is assigned the value of zero and a copy of the *this* register is made directly prior to calling CDialog::DoModal(). We came into this situation hoping to get through the code learning as little about MFC as we can. However, this code presents us with a little bit of a problem, because we're almost at the end of the code and it looks like it's all going to be MFC-centric from here on. MSDN tells us that DoModal() will create the dialog box that we just setup. So what we can expect to see are a series of call backs for handling the dialog. Let's fire up the debugger and single step in again.

After a lot of single-stepping, we finally get to a call in MFC that actually creates the window by calling CreateDialogIndirectParamA(). This API call sends a WM_INITDIALOG message to the dialog's appropriate procedure, thus this serves as our entry point into the GUI aspect of our application, which is the initialization callback for this dialog type.

Figure 8.20 Initialization Callback

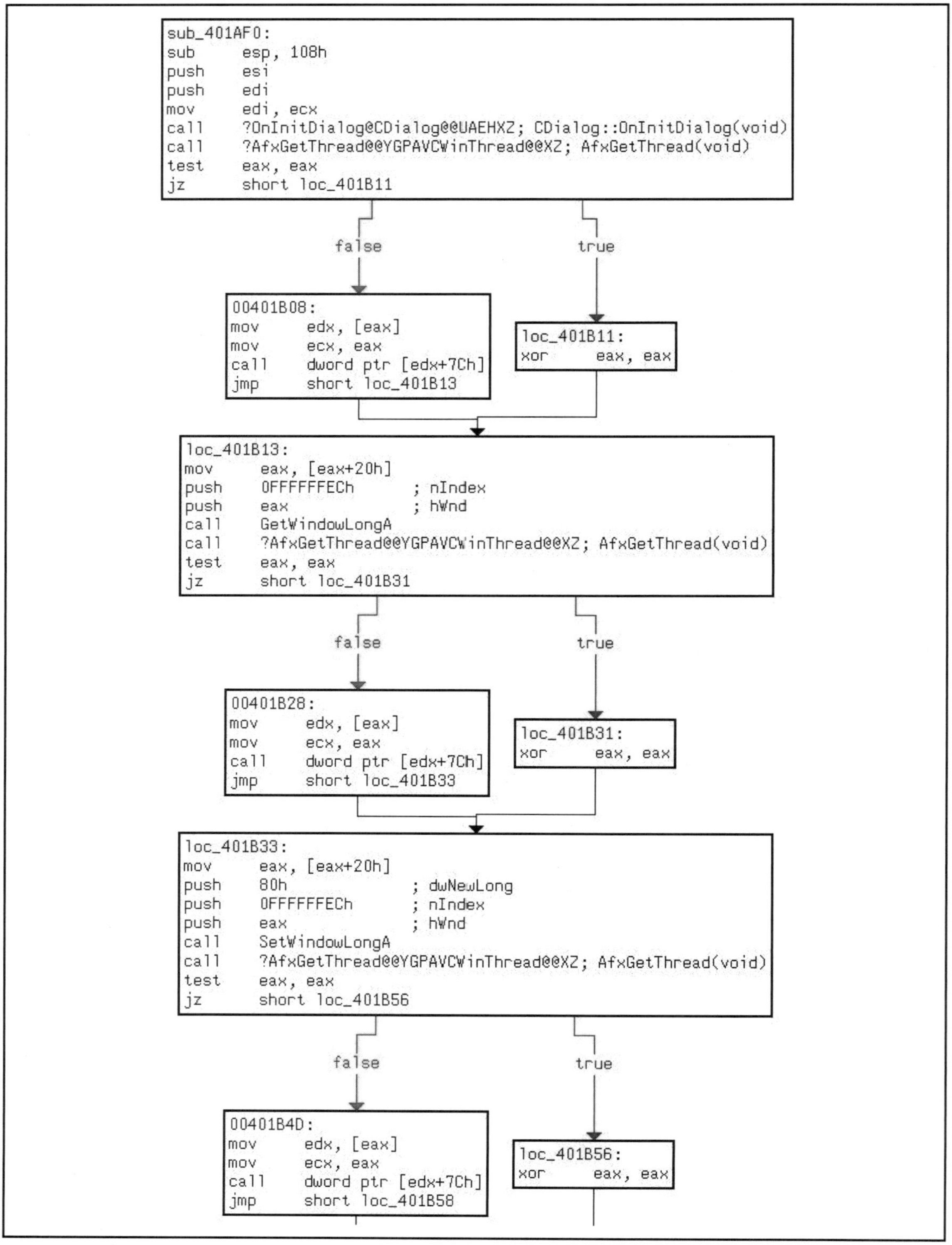

As we can see in Figure 8.20, the callback for initialization of the window is sub_401AF0(), which is another frameless method. Of what we see of the procedures initialization, there is not much that is interesting to us. We should single step through CDialog::OnInitDialog() to ensure that no derived procedures are called, and then also watch the indirect calls to EDX in the blocks labeled 0x00401B08, 0x00401B28 and 0x00401B4D. However, from what we can see thus far, it's mostly MFC code.

Stepping through we find the following in Figure 8.21. The call to CDialog::OnInitDialog() did not execute any user-defined code, and we can clearly see what the function pointer is in the following screenshot. It is important to note that because of the way that we're going through this code, and how fuzzy of a methodology we're employing, I've set breakpoints on and around the entry points for all functions defined in this binary. So at least in theory if we miss anything we will notice it and can back up to figure out what we missed, how we missed it and so on.

Figure 8.21

Moving into the next section of code in this function, we find the following code in Figure 8.22, which is a little more interesting and helps flesh out the purpose more.

Figure 8.22

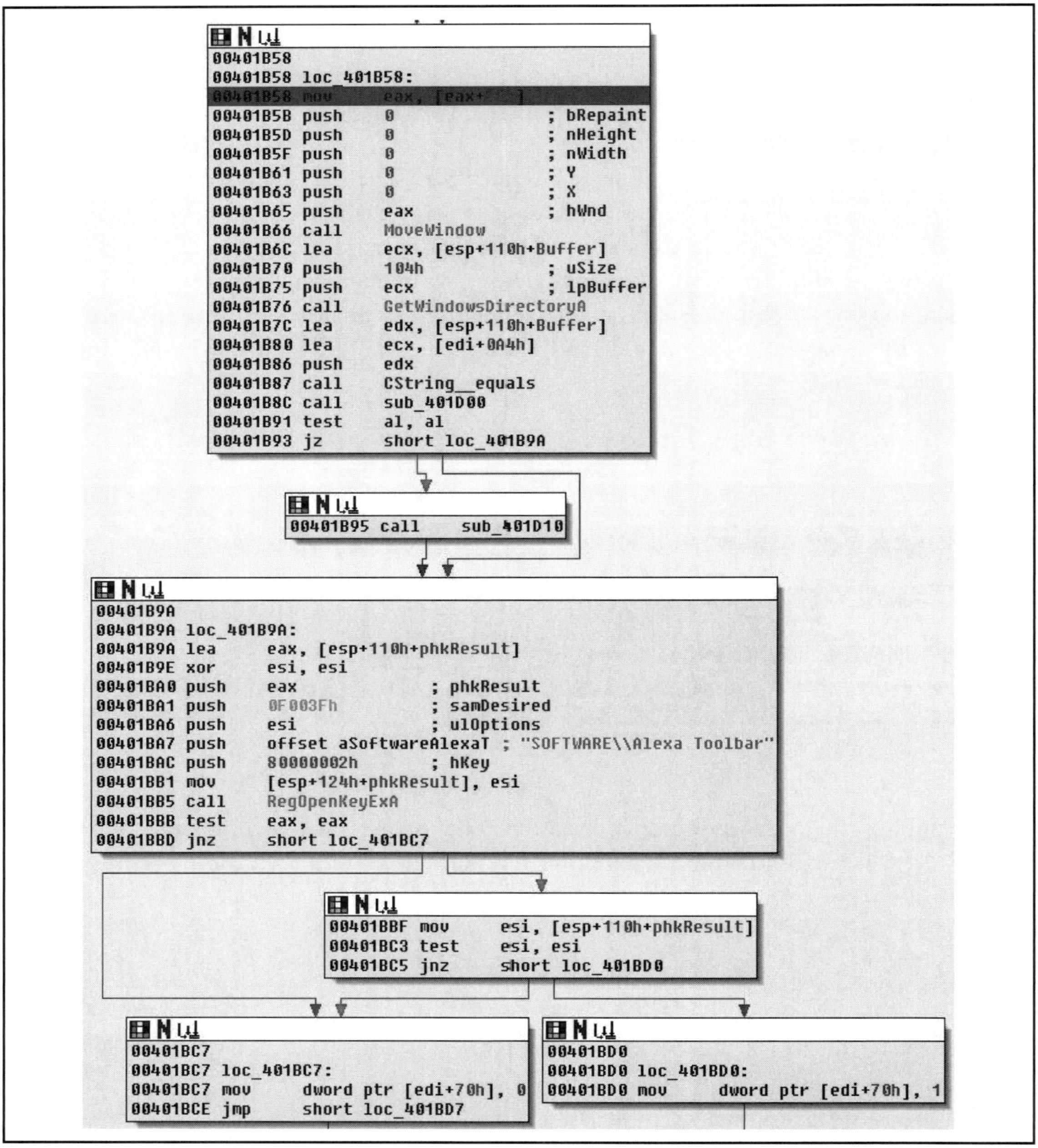

Here we find what we expected. Upon initialization of the window one of the first things done is a call to MoveWindow(), which will resize and/or reposition a window (Figure 8.23). In this instance they resize the window to be zero by zero at position zero by zero. From there we obtain the location of the windows directory via a call to GetWindowsDirectoryA() and assign the return value to the CString at EDI+0x0A4, followed by a call to sub_401D00().

Figure 8.23

```
00401D00
00401D00
00401D00
00401D00 sub_401D00 proc near
00401D00 call    GetVersion      ; Get current version number of Windows
00401D00                         ; and information about the operating system platform
00401D06 cmp     eax, 80000000h
00401D0B setnb   al
00401D0E retn
00401D0E sub_401D00 endp
00401D0E
```

As we can see, sub_401D00() is a simple procedure that just checks to see if the current version of Windows is at least Windows NT modifying the AL register based on this result and returning back to the caller. From there, we test the low-order 8-bits of the EAX register, either calling sub_401D10() or going to loc_401B9A() dependant on Windows version. If the version of the operating system is old enough, then all sub_401D10() does is call RegisterServiceProcess() to keep the application from exiting when the user logs off. From there, we see that it opens the registry key "Software\Alexa Toolbar" setting the variable at EDI+0x70 to 1 if the call succeeded and to zero if it did not.

Figure 8.24

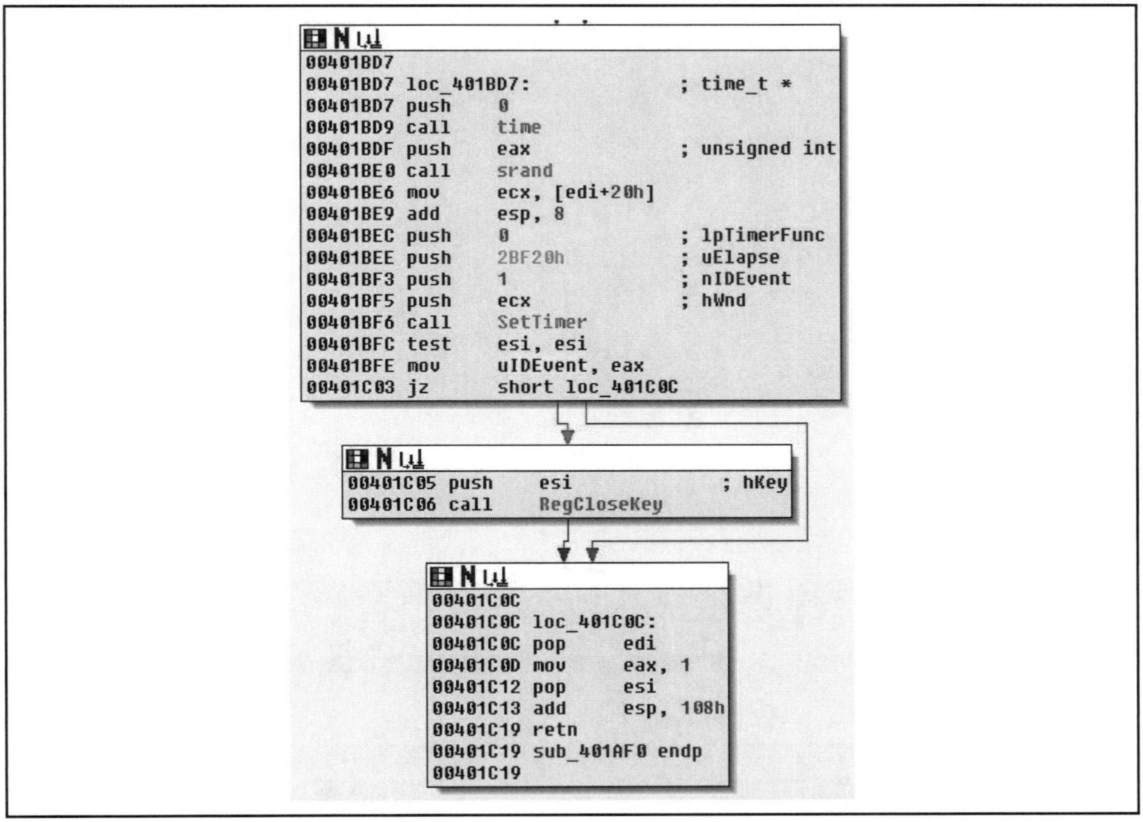

From there, we see that the pseudo-random number generator (PRNG) is seeded with the current time and then the SetTimer() API is called. Here we note that a NULL pointer is passed in for the callback parameter, which means that a WM_TIMER message will be sent to the window when the timer expires in 0x2BF20 milliseconds or three minutes roughly. From there, the registry key is closed and the routine returns. From there, we will single step some more and see where we end up, knowing that in about three minutes tops we will land at the procedure for the WM_TIMER handler.

Figure 8.25

```
mfc42.dll:69C5C860 loc_69C5C860:                     ; CODE XREF: mfc42.dll:69C5C8CB↓j
mfc42.dll:69C5C860 push    0
mfc42.dll:69C5C862 push    0
mfc42.dll:69C5C864 push    0
mfc42.dll:69C5C866 push    0
mfc42.dll:69C5C868 push    ebx
mfc42.dll:69C5C869 call    ds:PeekMessage
mfc42.dll:69C5C86F test    eax, eax
mfc42.dll:69C5C871 jnz     short loc_69C5C8D6
mfc42.dll:69C5C873 cmp     [ebp-4], eax
mfc42.dll:69C5C876 jz      short loc_69C5C88F
mfc42.dll:69C5C878 push    1
mfc42.dll:69C5C87A mov     ecx, esi
mfc42.dll:69C5C87C call    near ptr mfc42_6215
mfc42.dll:69C5C881 mov     ecx, esi
mfc42.dll:69C5C883 call    near ptr unk_69C5C206
mfc42.dll:69C5C888 mov     dword ptr [ebp-4], 0
mfc42.dll:69C5C88F
mfc42.dll:69C5C88F loc_69C5C88F:                     ; CODE XREF: mfc42.dll:69C5C876↑j
mfc42.dll:69C5C88F test    byte ptr [ebp+8], 1
mfc42.dll:69C5C893 jnz     short loc_69C5C8B1
mfc42.dll:69C5C895 mov     eax, [ebp-0Ch]
mfc42.dll:69C5C898 test    eax, eax
mfc42.dll:69C5C89A jz      short loc_69C5C8B1
mfc42.dll:69C5C89C test    edi, edi
mfc42.dll:69C5C89E jnz     short loc_69C5C8B1
mfc42.dll:69C5C8A0 mov     ecx, [esi+20h]
mfc42.dll:69C5C8A3 push    ecx
mfc42.dll:69C5C8A4 push    edi
mfc42.dll:69C5C8A5 push    121h
mfc42.dll:69C5C8AA push    eax
mfc42.dll:69C5C8AB call    ds:SendMessage
mfc42.dll:69C5C8B1
mfc42.dll:69C5C8B1 loc_69C5C8B1:                     ; CODE XREF: mfc42.dll:69C5C893↑j
mfc42.dll:69C5C8B1                                   ; mfc42.dll:69C5C89A↑j ...
mfc42.dll:69C5C8B1 test    byte ptr [ebp+8], 2
mfc42.dll:69C5C8B5 jnz     short loc_69C5C8CD
mfc42.dll:69C5C8B7 push    edi
mfc42.dll:69C5C8B8 push    0
mfc42.dll:69C5C8BA push    36Ah
mfc42.dll:69C5C8BF mov     ecx, esi
mfc42.dll:69C5C8C1 call    near ptr unk_69C3DA04
mfc42.dll:69C5C8C6 add     edi, 1
mfc42.dll:69C5C8C9 test    eax, eax
mfc42.dll:69C5C8CB jnz     short loc_69C5C860
mfc42.dll:69C5C8CD
mfc42.dll:69C5C8CD loc_69C5C8CD:                     ; CODE XREF: mfc42.dll:69C5C8B5↑j
mfc42.dll:69C5C8CD mov     dword ptr [ebp-8], 0
mfc42.dll:69C5C8D4 lea     ebx, [ebx]
mfc42.dll:69C5C8D6
mfc42.dll:69C5C8D6 loc_69C5C8D6:                     ; CODE XREF: mfc42.dll:69C5C85E↑j
mfc42.dll:69C5C8D6                                   ; mfc42.dll:69C5C871↑j ...
mfc42.dll:69C5C8D6 call    mfc42_1175
mfc42.dll:69C5C8DB mov     edx, [eax]
mfc42.dll:69C5C8DD mov     ecx, eax
mfc42.dll:69C5C8DF mov     eax, [edx+64h]
mfc42.dll:69C5C8E2 call    eax
mfc42.dll:69C5C8E4 test    eax, eax
mfc42.dll:69C5C8E6 jz      short loc_69C5C95E
mfc42.dll:69C5C8E8 cmp     dword ptr [ebp-4], 0
```

As we single-step past the dialog creation, we enter a section of code in the MFC DLL that acts as a message queue loop, processing windows messages. After the call to PeekMessage(), if there is a message in the queue then control is branched off to loc_69C5C8D6, and ultimately the call EAX instruction calls CWinThread::PumpMessage(). Eventually, after walking through the loop long enough we find the next callback, which is sub_401FC0() and is defined as shown in Figure 8.26.

Figure 8.26

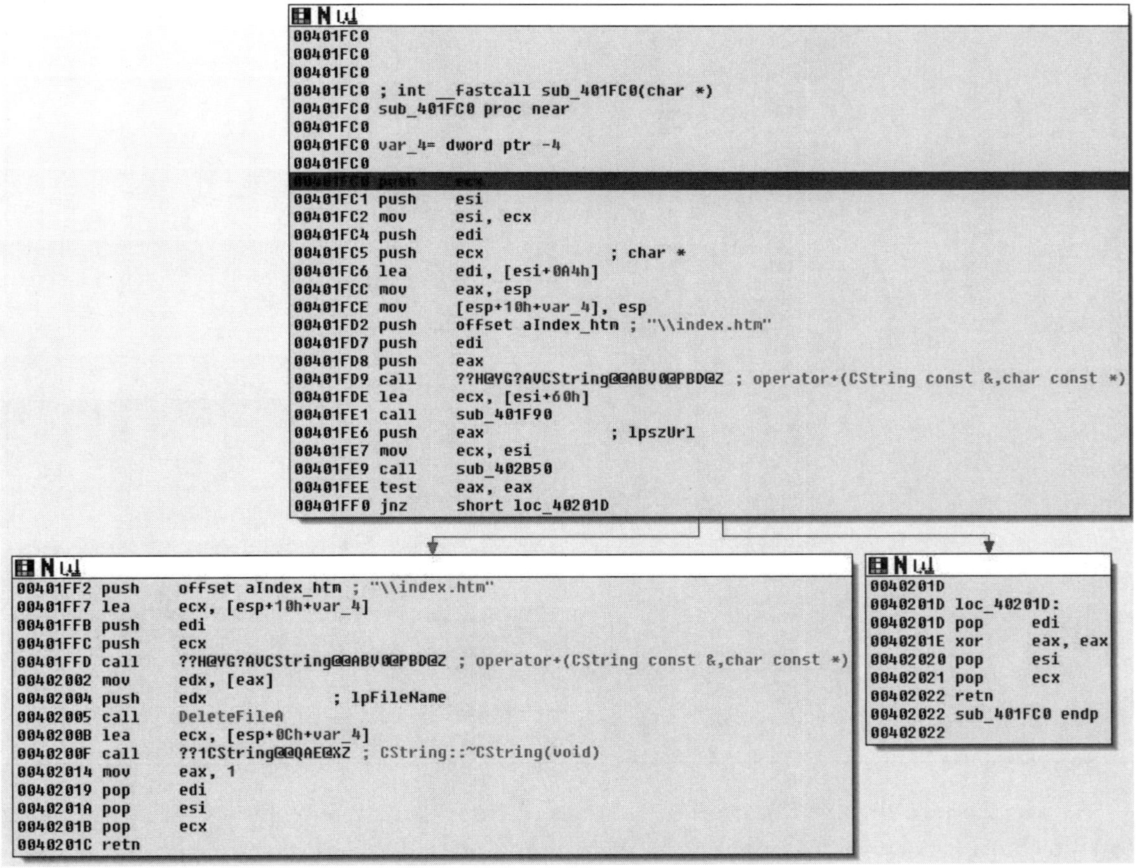

Upon entry into the function, we see a bit more that interests us. Specifically, one of the first things done is the string "\index.htm" and the Windows directory are concatenated together. This yields the string "C:\Windows\index.htm", which is followed by a call to the sub_401F90(). This simply dereferences the pointer in the ECX register and copies that pointer into the EAX register. Then a pointer to the string http://www.alxup.com/adsnt/AdsNT.ini and the newly constructed string are then passed as parameters to the routine sub_402B50() which we examine in Figure 8.27.

Figure 8.27

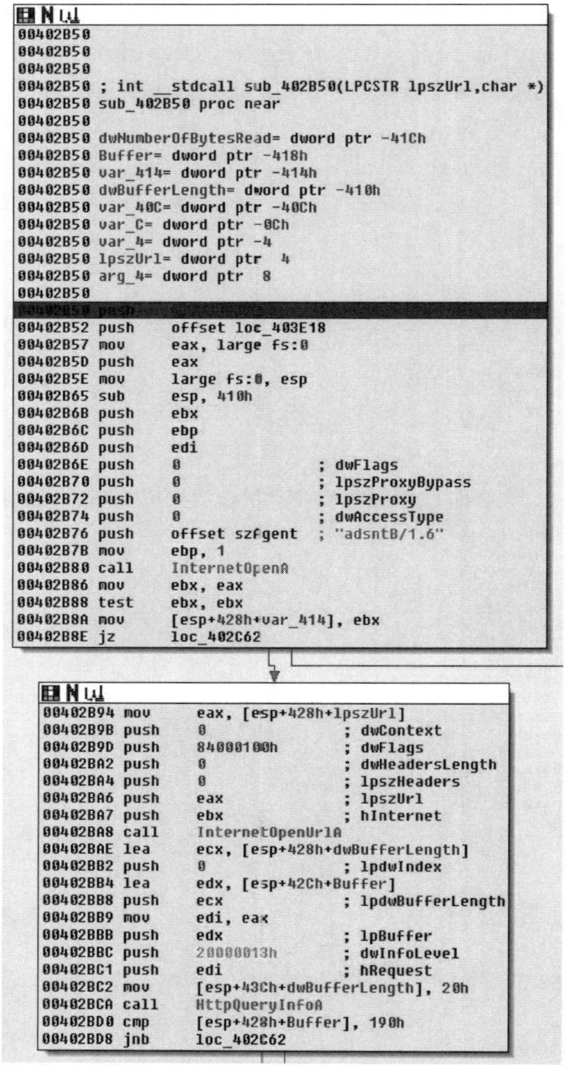

As we step into this new routine, it becomes obvious that we're really starting to get down to brass tacks with the application. We can now see it making contact with the outside world. After the procedure prologue we see a call to InternetOpenA(). The only interesting parameter to us in this instance is the User-Agent that will be employed. This gives us a static network detectable pattern to look for. For instance, suppose you wanted to generate an IDS rule, you would probably look for a packet that looks something like this (nuances of fragmentation/request splitting/weirdo tab placement/urgent data/et cetera ignored):

```
GET /adsnt/AdsNT.ini [...]
[...]
User-Agent: adsntB/1.6
[...]
```

The InternetOpenA() API call returns a handle, and we can see the return value copied into the EBX register. This is eventually copied into var_414, and renamed to 'inetHandle'. It should also be noted that the return value is checked against NULL. From there, the URL that was passed in as a parameter is retrieved via a call to InternetOpenUrlA(). After that has been retrieved, a call to HttpQueryInfoA() is made checking the status code returned by the remote web server. The code then checks this value against 0x190, which is 400 decimal and branches if the status was not less than 400. Now potentially (and I'm not positive because I would need to delve further into the InternetOpenURL() function than I am currently willing to), if the status code returned was larger than 0x20 in length than we would end up with an uninitialized variable, which could result in a bug. This is unlikely in this case because the flags passed to HttpQueryInfoA() specifies that we want the status code returned as an integer (0x20000013 is HTTP_QUERY_FLAG_NUMBER | HTTP_QUERY_STATUS_CODE). Moving on through the function body, we see Figure 8.28.

Figure 8.28

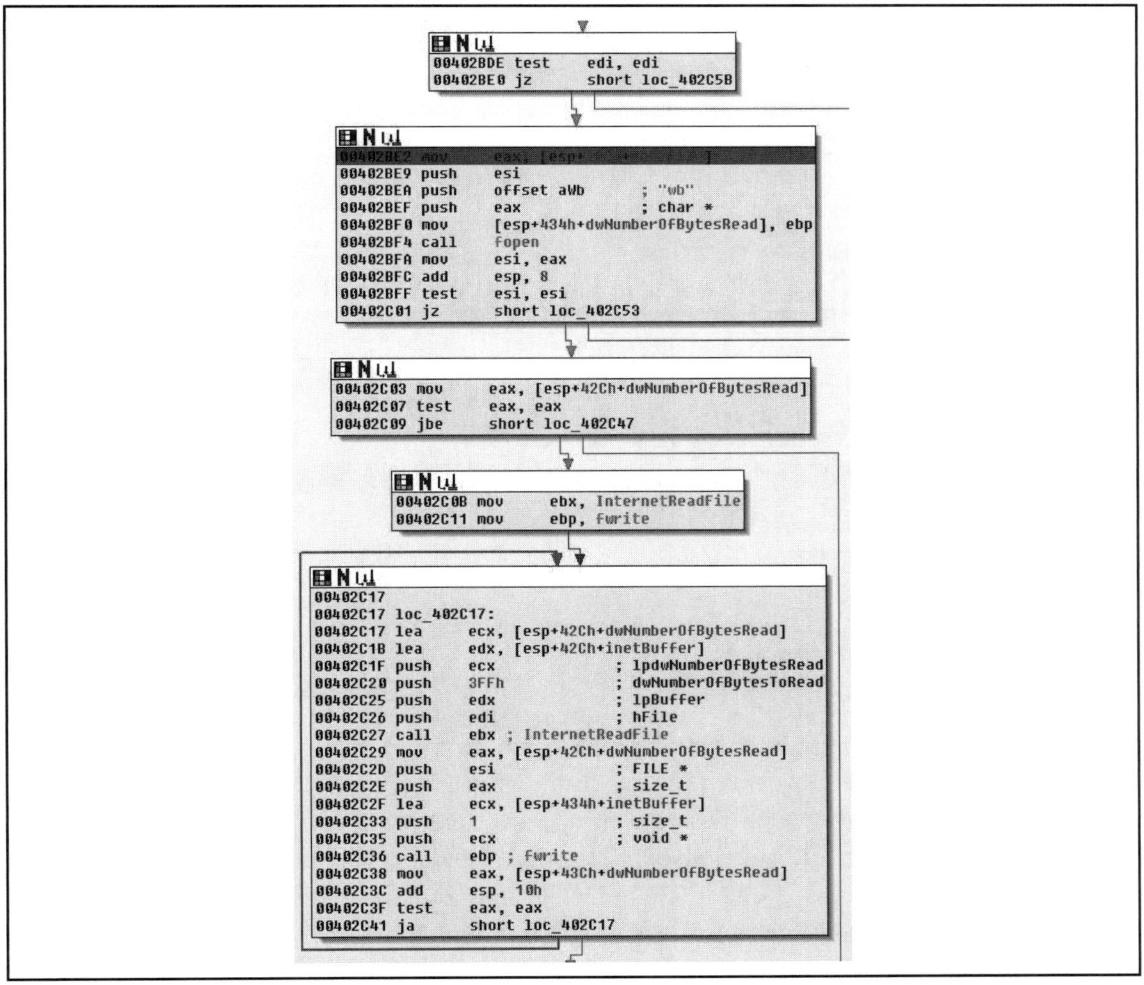

Moving down, the next thing we see is that the copy of the handle returned by InternetOpenA () is tested, branching if its NULL and otherwise moving on to a call to fopen(). Here we see the second parameter come into play as a file with that name is opened for binary writing. Next the number of bytes is checked, and if all is well we end up at loc_402C17. Here we see a simple loop where 0x3FF bytes at a time are read from the network and, then written to disk via calls to InternetReadFile() and fwrite() respectively. From this loop we enter the next section of code which is shown in Figure 8.29.

Figure 8.29

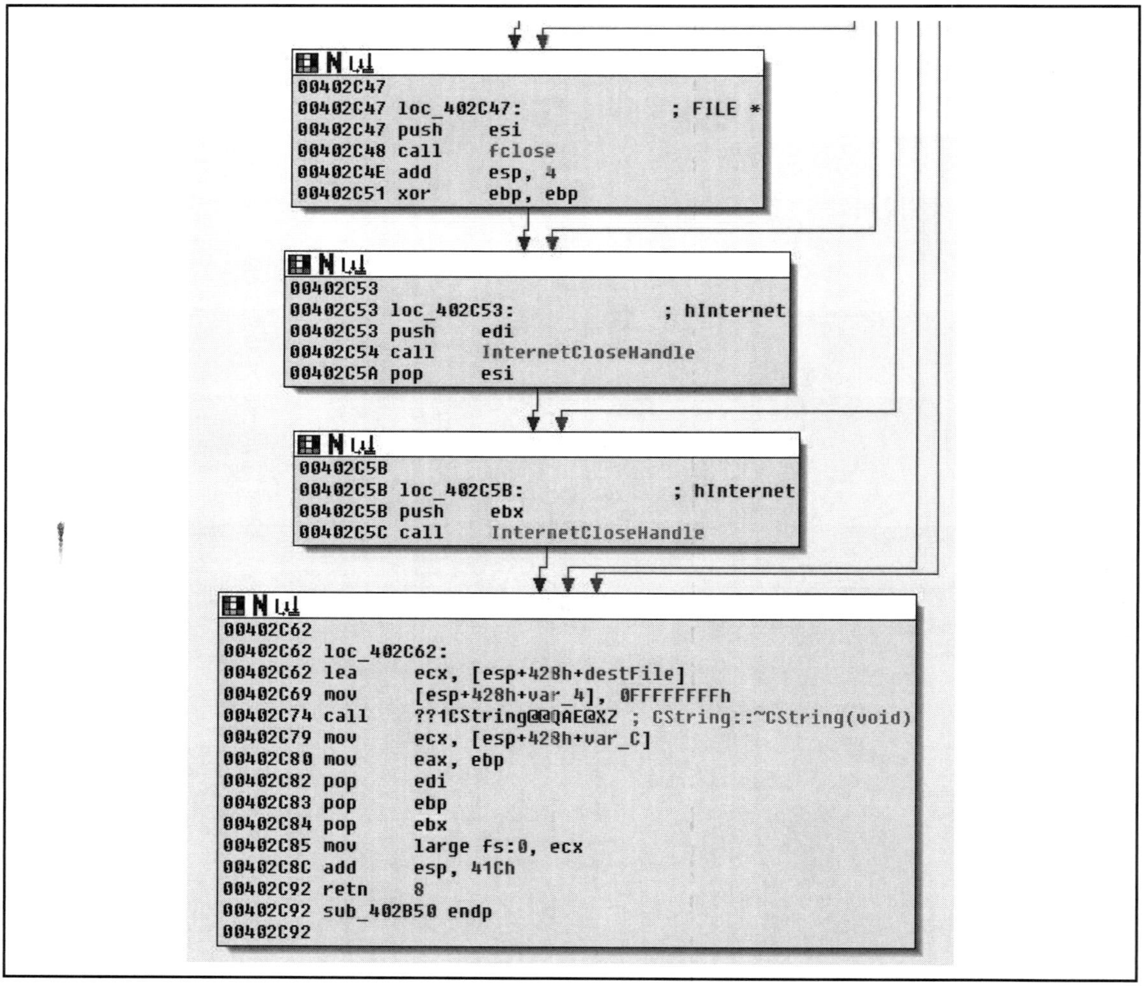

Moving into the final portion of the function, we see largely what we expected. Having read all of the file from the network and writing it to disk, we now close the file handle and internet connection, then call the destructor for the CString and return the EBP register as the return value. This is zeroed out at 0x00402C51, thus a successful I/O cycle in this routine returns a value of zero, and non-zero for failure. So in conclusion, this function takes two parameters: a URL and a path. The routine retrieves the file at the URL and saves it to the path specified. We will rename the function downloadToFile(). It should probably be noted that at least under Vista, there is some oddity that causes the file to be written to "C:\Users\[username]\AppData\Local\VirtualStore\" instead of C:\Windows. That all said, let's take a look at the INI file we downloaded.

```
[AdsNT]
Version=100
AdNum=9
[AdsNTURL]
leftpos1=-1000
toppos1=-1000
width1=1
height1=1
url1=http://www.deepdo.com/union/3721/yad.htm
objurl1=http://zzz.yy.xom
weight1=10
showIEWindow1=0
showStyle1=0
group1=0
leftpos2=-1000
toppos2=-1000
width2=1
height2=1
url2=http://talent.deepdo.com
objurl2=http://xx.you.com
weight2=100
showIEWindow2=0
showStyle2=0
group2=0
leftpos3=-1000
toppos3=-1000
width3=1
height3=1
url3=http://www.deepdo.com/union/3721/yad.htm
objurl3=http://xxx.yyy.com/
weight3=10
showIEWindow3=0
showStyle3=0
group3=0
leftpos4=-1000
```

```
toppos4=-1000
width4=1
height4=1
url4=http://www.92site.cn/search/135go.jsp
objurl4=http://xxx.com
weight4=15
showIEWindow4=0
showStyle4=0
group4=0
leftpos5=-1000
toppos5=-1000
width5=1
height5=1
url5=http://www.5isou.cn/
objurl5=http://xxx.xxx.com/
weight5=100
showIEWindow5=0
showStyle5=0
group5=0
leftpos6=-1000
toppos6=-1000
width6=1
height6=1
url6=http://www.deepdo.com/site.htm
objurl6=http://xxx.xxx.com/
weight6=150
showIEWindow6=0
showStyle6=0
group6=0
leftpos7=-1000
toppos7=-1000
width7=1
height7=1
url7=http://www.5isou.cn/calendar/index.htm
objurl7=http://xxx.xxx.com/
weight7=150
showIEWindow7=0
showStyle7=0
```

```
group7=0
leftpos8=-1000
toppos8=-1000
width8=1
height8=1
url8=http://www.92site.cn/search/135go.jsp
objurl8=http://xxx.xxx.com/
weight8=150
showIEWindow8=0
showStyle8=0
group8=0
leftpos9=-1000
toppos9=-1000
width9=1
height9=1
url9=http://www.iesafe.cn/yahoo/index.htm
objurl9=http://xxx.xxx.com/
weight9=150
showIEWindow9=0
showStyle9=0
group9=0
[AdsNTGroupURL]
gurl21=http://www.deepdo.com/union/iplus/edodo.htm
```

As you can see, in our file we have a list of URLs and a series of positions and sizes; we can probably safely assume then that we're looking at ad-ware, and that the sizes/positions refer to browser window placement. What's annoying, though, is that the application we're going to download, probably downloads some other application, which downloads another and so on. Even more, we can bet that many of these URLs also host more malicious content. It's really a losing battle for the user who made the mistake of clicking something they didn't mean to once as their computer ends up overloaded. This is interesting because if you look at many viruses and worms, the idea of coexistence with the host platform is something that can often be found, to the point that there are actually viruses that attempt to remove other viruses that they recognize. Adware and Spyware are the exact opposite. I suppose it could be argued that the motives are different and that greed causes the corporate malware to take the approach of infecting the computer to the point of it being useless as they don't expect to survive for any period of time anyways. Nonetheless, enough with the theory, let's get back to the application shall we?

Moving back to the calling function we have the code in Figure 8.30 left to analyze.

Figure 8.30

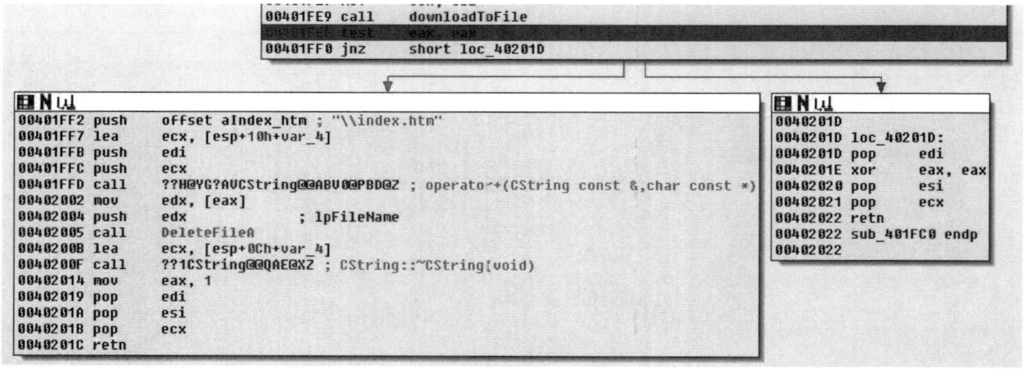

So, after the return value is checked and the call to downloadToFile() succeeded, we continue execution at 0x00401FF2, or loc_40201D if the call failed. The CString for the path is once again constructed and a call to DeleteFileA() is called, removing the file. After that the destructor for the CString is called and we return to the caller, which the message handling code for the application. From there we enter the next function, sub_401DF0(), which is defined as shown in Figure 8.31.

Figure 8.31

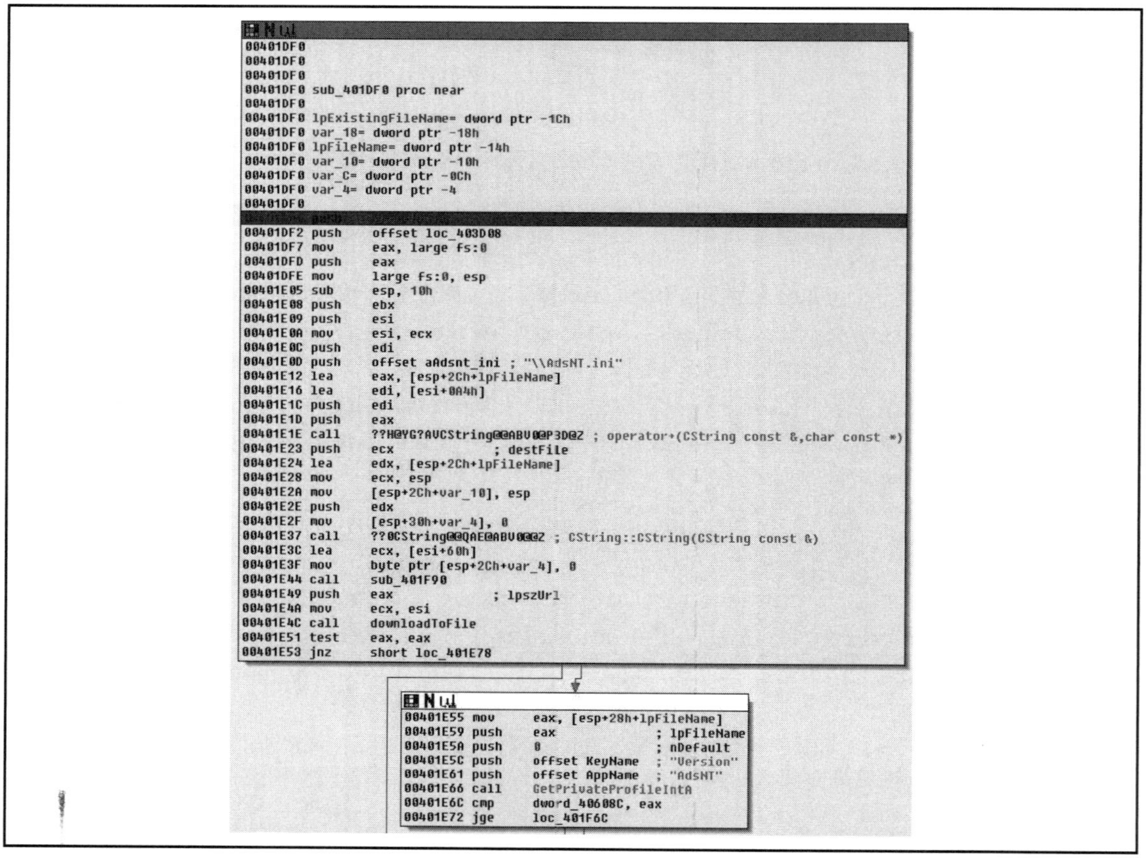

Upon entering into this new subroutine, we first see that the path to the AdsNT.ini file is again constructed and then downloaded, branching off to loc_401E78 if there was an error, otherwise continuing at 0x00401E55. From there, we see that the Version is retrieved from the INI file via a call to GetPrivateProfileIntA() and then a conditional branch based off of that, thus we have a simple version compatibility check. Moving on to the next section, we see the code from Figure 8.32.

Figure 8.32

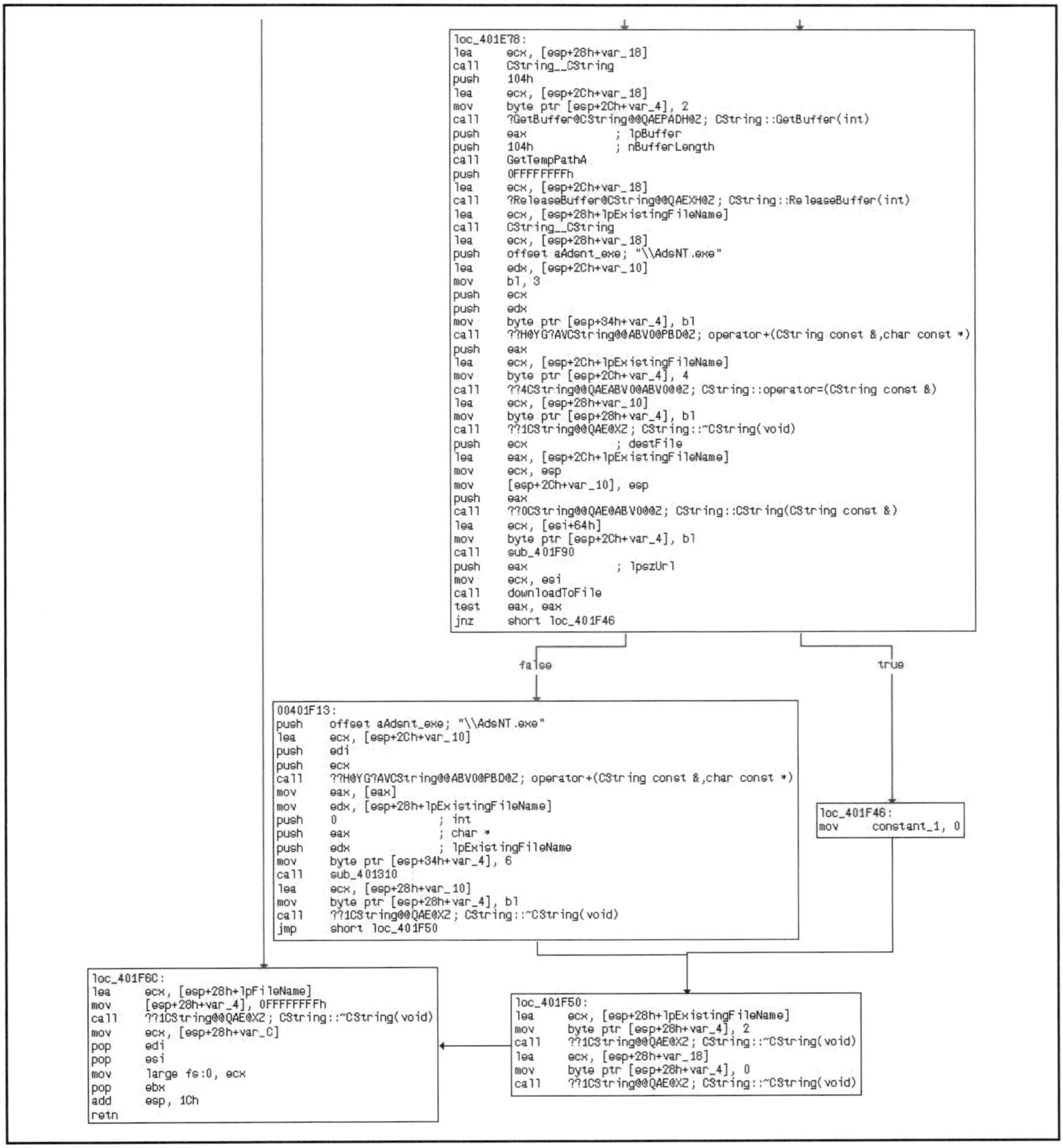

Starting at loc_401E78, we see a fairly familiar pattern, we see a series of CString's being created, although we see a slight twist in that a temporary filename is retrieved via a call to GetTempPathA(). From this path we ultimately note that a temporary filename and the string "\AdsNT.exe" are concatenated together. From there, we can see exactly what is going on by the call to downloadToFile(), which if you remember was the routine we reversed immediately prior to this routine and that simply acted as a go between for a URL and a local file on the disk, fetching a remote file and saving it. Thus the actual executable image that came up at the first object creation is finally being downloaded.

Thus far we have concluded that the application is definitely malicious in the sense that it is obviously adware. It was packed and attempted to obfuscate the strings stored internally, the later especially should raise red flags for you. Furthermore, an INI file was downloaded that after review it became quite clear that it contained information such as URLs and coordinate and window sizes presumably for those URLs. Things look pretty suspicious, but there is still a lot of work to be done. This is okay though, as it provides an excellent opportunity for you, the reader, to apply the knowledge learned in this book. On the Syngress website you can find a copy of this program and repeat some of the steps we have, such as unpacking the application. From there, consider the following exercises:

Exercise 0: Starting at 0x00401F13, reverse the rest of the application-how does the rest of it operate?

Exercise 1: Reverse/Analyze AdsNT.exe

Exercise 2: In dealing with the message queue, we skipped over a section of code that is actually defined by the application and serves as the callback that ultimately calls the majority of the functions we've analyzed here. Starting at the end of the first routine to call downloadToFile(), the one that first obtained the INI file, walk through until the routine returns and make note of where in the application control flow is handed to, not step backwards through the application until the initial entry point is realized.

Chapter 9

IDA Scripting and Plug-ins

Solutions in this chapter:

- Introduction
- Basics of IDA Scripting
- IDC Syntax
- Simple Script Examples
- Writing IDC Scripts
- Basics of IDA Plug-ins
- Plug-in Syntax
- Simple Plug-in Examples
- The Indirect Call Plug-in

☑ Frequently Asked Questions

Introduction

IDA Pro is a tool used by many people in different areas of reverse engineering. The user base includes malware analysists, vulnerability researches, software reversers, hackers, firmware/hardware reversers, developers, software protection enthusiasts, and many more.

IDA Pro's extensibility is what makes it a great tool. The interactive nature of IDA goes very well with scripting and writing plug-ins. IDA is a tool in the true sense of the word. The user guides IDA to achieve what is needed.

This chapter is about extending IDA Pro with scripts or plug-ins. As we reverse engineer binaries, eventually we start seeing patterns. These patterns can be code patterns or repetitive tasks that are ripe for automation.

IDA can be extended using various methods. IDC is the built in scripting language. IDC is C-like in structure and since it is interpreted, no other tools are needed. More complicated tasks are relegated to plug-ins. Hex-rays provides an SDK to customers allowing for plug-in development. The SDK is written in C++ with support for various compilers. Third party hybrid solutions have also been developed. These hybrid solutions wrap IDC functions as well as some SDK functions. (You can download code and scripts in this chapter from www.syngress.com/solutions).

Basics of IDA Scripting

IDC is IDA Pro's built in scripting language. It is very similar to C in syntax. Someone familiar with C should be able to pick up IDC quickly. It is interpreted.

There are two standard ways to execute IDC.

- IDC Statements can be executed directly from within IDA. **SHIFT+F2** brings up a dialog box. Statements entered in the box will be executed. The dialog box is used to enter in small code snippets. Functions are generally not defined in the dialog box, although there is a w
- IDC files can be loaded. To load an IDC file go to **File | IDC File**. A file browse dialog will come up.

TIP

Another option to execute IDC expressions is through an optional command line. The command line option must be set to yes in idagui.cfg.

DISPLAY_COMMAND_LINE = YES // Display the expressions/IDC command line.

IDC Syntax

The IDC scripting language borrows a great deal of syntax from C. All statements end with a semicolon. The similarity includes many of the same keywords including if, if else, while, do while, continue and break. This section will introduce IDC syntax highlighting differences between IDC and C.

Scripting provides access to the disassembly with much less effort than writing plug-ins. Scripts can be run from files as well as the IDC dialog box. The examples in this section will use the dialog box until we reach functions. The use of even simple scripting will speed up analysis and help automate tasks. In order to run IDC scripts, an idb file must be loaded into IDA.

Output

The first thing taught in most programming books, since K&R C, is the hello world program. Getting data out to the user from a script is very important. This can be actual output or even just for debugging purposes.

Open the IDC command window by pressing **SHIFT+F2** or using the menus (**File | IDC** Command…) and type:

```
Message("Hello world\n");
```

The dialog should appear similar to Figure 9.1. Multiple statements can be entered, but for now just the *Message* statement will suffice. After clicking **OK**, *hello world* should appear in the message window.

Figure 9.1 Hello world

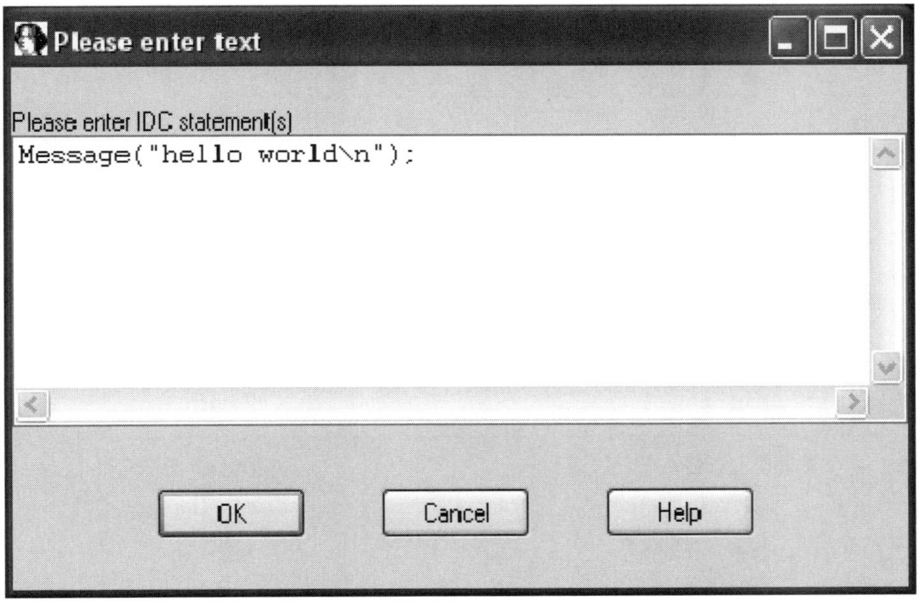

The Message function is similar to the C *printf* function, also using format strings. The function prototype is:

```
void Message (string format,…);
```

Some other variants using *Message*:

```
Message("%s\n", "hello world");
Message("%x\n", 0x40100);
Message("%x is the cursor\'s address\n", ScreenEA());
```

The first example uses the "%s" format string also printing *hello world*. The second example uses "%x" to print out a hexadecimal value. While the other two examples are somewhat contrived, the third one uses a new IDC function. The *ScreenEA* function returns the current cursor address. This function is commonly found in scripts.

Message is not the only output, but it is the most commonly used. Two other functions are available, *Warning* and *Fatal*. Both of these use format strings and have the same function prototype.

Warning is used to alert the user of problems. It will bring up a box similar to Figure 9.2. *Fatal* is rarely used since it terminates IDA without saving the database.

Figure 9.2 Warning Box

Variables

All variables in IDC are defined using the auto type. The statement below declares a variable called counter.

```
auto counter;
```

The auto type can represent	Example
32bit integer (64 bit in IDA Pro 64)	0x00401000
character string	"hello world"
floating point number	5.23

Variables have size limits depending on the type of data they contain.

- Integers are 32 bit (64 bit for IDA Pro 64).
- Character strings can be up to 1023 characters long.
- Floating point variables are up to 25 decimal digits.

An auto variable can represent different types of data. As such, there are conversion rules when operating on different types. Generally when scripting, type conversions are not as common as in C. There are functions to manually perform type conversion:

```
long(expr)
char(expr)
float(expr)
```

Variables must be declared and assigned in separate statements.

```
auto currentAddress;
currentAddress=ScreenEA();
```

Most of the standard C operators (+, −, /, *, %, <<, >>, ++, −−) work in IDC. Some operators are unsupported, these include the combination assignment operators (+=) and comma operation (,). Unlike C, added strings will be concatenated.

All variables have local scope. This means they are only available within the function that defined them. For our current purposes, this applies to the IDC command window. Functions are covered later as well as way to allow global variables.

Conditionals

Most of the standard C conditional statements are available. These include *if*, *if else*, and the ternary operator "? :". The switch statement is not available in C. This code snippet shows *if else*.

```
auto currAddr;
currAddr = ScreenEA();
if (currAddr % 2)
   Message("%x is odd\n", currAddr);
else
   Message("%x is even\n", currAddr);
```

Loops

Looping can be done by for, *while*, and *do while*. These are similar to C except the comma operator is not allowed. The switch statement is not supported in IDC, but multiple if, else if statements can be used. The code snippet demonstrates a loop and introduces some new IDC functions and concepts.

```
auto origEA, currEA, funcStart, funcEnd;

origEA = ScreenEA();
funcStart = GetFunctionAttr(origEA, FUNCATTR_START);
funcEnd = GetFunctionAttr(origEA, FUNCATTR_END);

if(funcStart == -1)
   Message("%x is not part of a function\n", origEA);
for(currEA=funcStart; currEA != BADADDR;currEA=NextHead(currEA, funcEnd))
{
    Message("%8x\n", currEA);
}
```

NOTE

BADADDR is a constant used in IDC. It represents an error or invalid result from function. In scripts it is used to test results and sometimes initially assigned to variables.

Some IDC functions return –1 on an error. Internally BADADDR is represented by –1.

The code snippet prints out the addresses to every instruction within the current function. It introduces two new IDC functions, *GetFunctionAttr* and NextHead. The prototypes are:

```
long GetFunctionAttr(long ea, long attr);
long NextHead(long ea, long maxea);
```

GetFunctionAttr allows us to query for certain function attributes. The argument ea is any address within the function. The argument attr is the specific attribute we are interested in. In this case we are looking for a function start and end given an address. If the address *ea* is not within a function then GetFunctionAttr returns −1.

NextHead returns the next instruction or data item. The argument *ea* is the start address and *maxea* is the end address. If there are no defined instructions or data in the given address range, then *BADADDR* is returned. On architectures like the IA-32 containing variable length instructions *NextHead* must be used to iterate through instructions. On RISC architectures with set length instructions one may be tempted to simply increment instead of using NextHead. This should be avoided as simply incrementing will not check if the item has been defined by IDA.

The code snippet demonstrates the same functionality using a *while* loop.

```
auto origEA, currEA, funcStart, funcEnd;

origEA = ScreenEA();
```

```
funcStart = GetFunctionAttr(origEA, FUNCATTR_START);
funcEnd = GetFunctionAttr(origEA, FUNCATTR_END);
if (funcStart == -1)
    Message("%x is not part of a function\n", origEA);
currEA = funcStart;
while (currEA != BADADDR)
{
       Message("%8x\n", currEA);
       currEA = NextHead(currEA, funcEnd);
}
```

Functions

Functions are needed once we move on from simple snippets of IDC. Functions are also required within IDC files. All functions in IDC are defined as static. The code below is an example function.

```
static outputCurrentAddress(myString)
{
     auto currAddress;
     currAddress = ScreenEA();
     Message("%x %s\n", currAddress, myString);
     return currAddress;
}
```

Function declarations in IDC have a few differences with C. The differences relate to types. Since IDC has only one type, *auto*, types are not needed in the arguments or return. IDA only accepts functions without types in the declaration. IDC files will be covered in the "Simple Script Examples" section later in this chapter.

Usually functions are only declared in an IDC file. Entering the above function in the IDC Command window produces an error "Syntax error near static". This is due to how the IDC Command window operates. A solution was first documented by Willem Jan Hengeveld (http://www.xs4all.nl/~itsme/projects/disassemblers/ida.html).

Internally the dialog box contents are stored in a function called *_idc*. Thus entering in a function declaration is actually attempting to declare a function within another function. The *_idc* function needs to be closed before a new function is declared. The new function must also leave off its closing brace. In order to declare *outputCurrentAddress* enter:

```
}
static outputCurrentAddress(myString)
{
     auto currAddress;
     currAddress = ScreenEA();
```

```
    Message("%x %s\n", currAddress, myString);
    return currAddress;
```

We should not see an error. Although we declare the function it does not execute. If we want to execute something as well, it needs to be part of the *_idc* function.

```
    AddHotkey("Alt-f9", "outputCurrentAddress2");
    outputCurrentAddress2();
}
static outputCurrentAddress2()
{
    auto currAddress;
    currAddress = ScreenEA();
    Message("%x\n", currAddress);
    return currAddress;
```

The preceding code introduces a new IDC function, *AddHotKey*. This function binds a key combination to an IDC function name. The target function can not have arguments. The hotkey is added and outputCurrentAddress2 is executed. The function *outputCurrentAddress2* can be executed via the hotkey or a call from the command window.

Local and Global Scope

Scope is what variables or functions are visible from a certain location in the code. We will use the variable *currAddress* from the *outputCurrentAddress* as an example of *local scope*. The *currAddress* variable is only visible from within its function. It cannot be accessed from another function.

Function declarations are placed in the *global scope*. Functions can be called from other functions. This includes calling from the command window. Once a function is defined it remains in the global scope until we either declare another function with the same name or until the IDA session is terminated. Closing an *idb* file clears out any IDC functions from memory.

Once the *outputCurrentAddress* is declared we can call it by entering into the command box.

```
outputCurrentAddress("some string");
```

We can have our own library of IDC functions load with IDA by adding them to ida.idc file. This file is located in the idc directory within the IDA Pro install directory.

TIP

Consider using a custom prefix to avoid name conflicts with functions from other scripts.

Global Variables

Auto variables are only in scope for the function they are defined. We need a way to have persistent data throughout our script. We need global variables. IDC does not provide direct support for global variables. However, global variables can be simulated using arrays.

Arrays are built in to IDC. An array can contain either string data or a long. The following code will create an array and define some items.

```
auto gArray;
gArray = CreateArray("myGlobals");
```

The code introduces a new IDC function, *CreateArray*. The prototype for *CreateArray* is:

```
long CreateArray(string name);
```

The name must be less than 120 characters. The function will either return the array id on success or −1 if the array creation fails. The following code adds some items to the array.

```
SetArrayLong(gArray, 23, 415);
SetArrayString(gArray, 0, "some string data");
```

The prototypes for these new functions are:

```
/*
arguments:
    id       -    array id
    idx      -    index of an element
    value    -    32bit value to store in the array
    str      -    string to store in array element
returns: 1-ok, 0-failed
*/
success SetArrayLong(long id,long idx,long value);
success SetArrayString(long id,long idx,string str);
```

The index *idx* can be any 32bit number. There is no need to use sequential indexes as space is only allocated as it is assigned. The previous example assigns the value 415 to index 23 and the string "some string data" to index 0. The global data is now assigned and can be accessed from anywhere else in the script, other scripts, or through the command window.

In order to access the global data, new IDC functions are introduced. The id of the array is needed to access its members. The following code demonstrates the new IDC functions.

```
auto arrayId, strItem, longItem;
arrayId = GetArrayId("myGlobals");
strItem = GetArrayElement(AR_STR, id, 0);
longItem = GetArrayElement(AR_LONG, id, 23);
```

Two new IDC functions are introduced, *GetArrayId* and *GetArrayElement*. The prototypes for the new functions are:

```
// get array id by its name
// arguments: name - name of existing array.
// returns:        -1 - no such array
//                      otherwise returns id of the array

long GetArrayId(string name);

/* get value of array element
        arguments: tag      - tag of array, specifies one of two
                                array types:
#define AR_LONG 'A' // array of longs
#define AR_STR 'S' // array of strings
                          id       - array id
                          idx      - index of an element
       returns:           value of the specified array element.
                          note that this function may return char or long
                          result. Unexistent array elements give zero as
                          a result.
*/
string or long GetArrayElement(long tag,long id,long idx);
```

The order of *GetArrayElement*'s arguments is different than the *SetArray* functions. These functions can be used anywhere after the array has been defined. The above code snippets do not have any error detection for space purposes, but error checks should be added.

Some IDC libraries of commonly used functions have been released. These libraries include global variables among other things and check for errors. One of these, *common.idc*, is written by lallous http://www.openrce.org/downloads/details/81/Common_Scripts. It includes other useful functions besides global variables. The following snippet is an example of using global variables with the common.*idc helper* functions.

First we need to initialize the global variables, using *InitGlobalVars*, mostly likely this will appear in the main function of the script.

```
if (InitGlobalVars() == 0)
{
    Message("InitGlobalVars() failed\n");
}
```

Once initialized, we have access to four macro definitions for writing and reading global variables. The following are the macros using the same naming convention as used earlier (index, value, string).

```
SetGlobalVarLong(index, value)
SetGlobalVarString(index, string)
GetGlobalVarLong(index)
GetGlobalVarString(index)
```

Setting array elements using the same data as the previous example:

```
SetGlobalVarLong(23, 415)
SetGlobalVarString(0, "some string data")
```

Accessing the items is also much cleaner and simpler.

```
auto strItem, longItem;

strItem = GetGlobalVarString(0)
longItem = GetGlobalVarLong(23)
```

Global variables are very useful as persistent information is often needed. The library can be added to *idc.idc*, allowing access from all scripts. Any functions we find useful can be added as well.

TIP

Messages should contain the address of interest as the leftmost item. For example:

```
Message("%x breakpoint set\n", bpAddr);
4014c6 breakpoint hit
```

This allows the address to be double-clicked taking IDA to the address.

Simple Script Examples

So far most of the examples have been using the IDC command window. The command window is great for interactive scripting, but it soon becomes unwieldy. Scripts allow us to run IDC code without having to re-enter it into a dialog window.

What are the differences between code snippets and scripts? There aren't many differences. All code must reside within functions. Even the command window code was located in the *_idc* function. The following outlines a script.

```
#include <idc.idc>

static some_function()
{

}
```

```
static main()
{
}
```

IDC use preprocessor directives like C. The file *idc.idc* contains IDC function prototypes and constants and it is usually included in all scripts. The file also serves as documentation, it contains the same information as the help file. IDC supports the *#define*, *#ifdef*, and other command preprocessor commands.

The main function is executed by the script. If a main function is not included, the other functions will remain in the memory and still be callable.

Tools & Traps…

Setting up an IDC Development Environment

There are various things that we can do to facilitate developing IDC scripts. A proper text editor is very important.

Our text editor should support syntax highlighting. Syntax highlighting uses different colors, font, and font sizes to represent different types of data. This allows us to easily identify IDC function calls and keywords from other data. Most modern editors feature syntax highlighting. Your favorite text editor most likely has an option to add new syntaxes. Sebastian Porst posted an IDC syntax file for the Crimson text editor http://www.the-interweb.com/serendipity/exit.php?url_id=157&entry_id=26.

Along with the proper editor, I change the file extension *idc* to open with the editor. IDA has option to set an editor of your choice for editing IDC scripts. This option is available from **Options | General | Misc | Editor**.

IDA by default will browse for IDC files from the last location opened. This behavior can be changed by editing a configuration file. The cfg directory within the IDA install directory holds many configuration files. We are particularly interested in *idagui.cfg*, which has options relating to idag.exe.

When developing and using IDC files, a commonly changed option is *OPEN_DEFAULT_IDC_PATH*. By default this option is set to **NO**. Changing the option to **YES**, will always open the IDC file browse dialog in the *idc* directory. A restart is required.

The script from Figure 9.3 will reset a function back to the default color. A coverage tool will color basic blocks as it traces execution, similar to Figure 9.4. Other times a user will color blocks to highlight certain code. In either case whether between tracing runs or if we are no longer interested in certain, we will need to reset the colors.

Figure 9.3 ResetColor IDC Script

```
#include <idc.idc>
static main(void)
{
    auto origEA, currEA, currColor, funcStart, funcEnd;
    origEA = ScreenEA();
    funcStart = GetFunctionAttr(origEA, FUNCATTR_START);
    funcEnd = GetFunctionAttr(origEA, FUNCATTR_END);

    Message("Welcome to resetColor.idc\n");
    if (funcStart == -1 || funcEnd == -1)
    {
        Message("** Error: not in a function **\n");
        return -1;
    }
    Message("[*] Function: %s\n", GetFunctionName(funcStart));
    Message("[*] start == 0x%x, end == 0x%x\n", funcStart, funcEnd);
    for (currEA = funcStart; currEA != BADADDR; currEA =
NextHead(currEA, funcEnd))
    {
            if (SetColor(currEA, CIC_ITEM, DEFCOLOR) == 0)
            {
                Message("** Error: SetColor failed 0x%x **\n", currEA);
            }
    }
    Refresh();
    Message("resetColor is done\n");
}
```

Figure 9.4 Traced Function

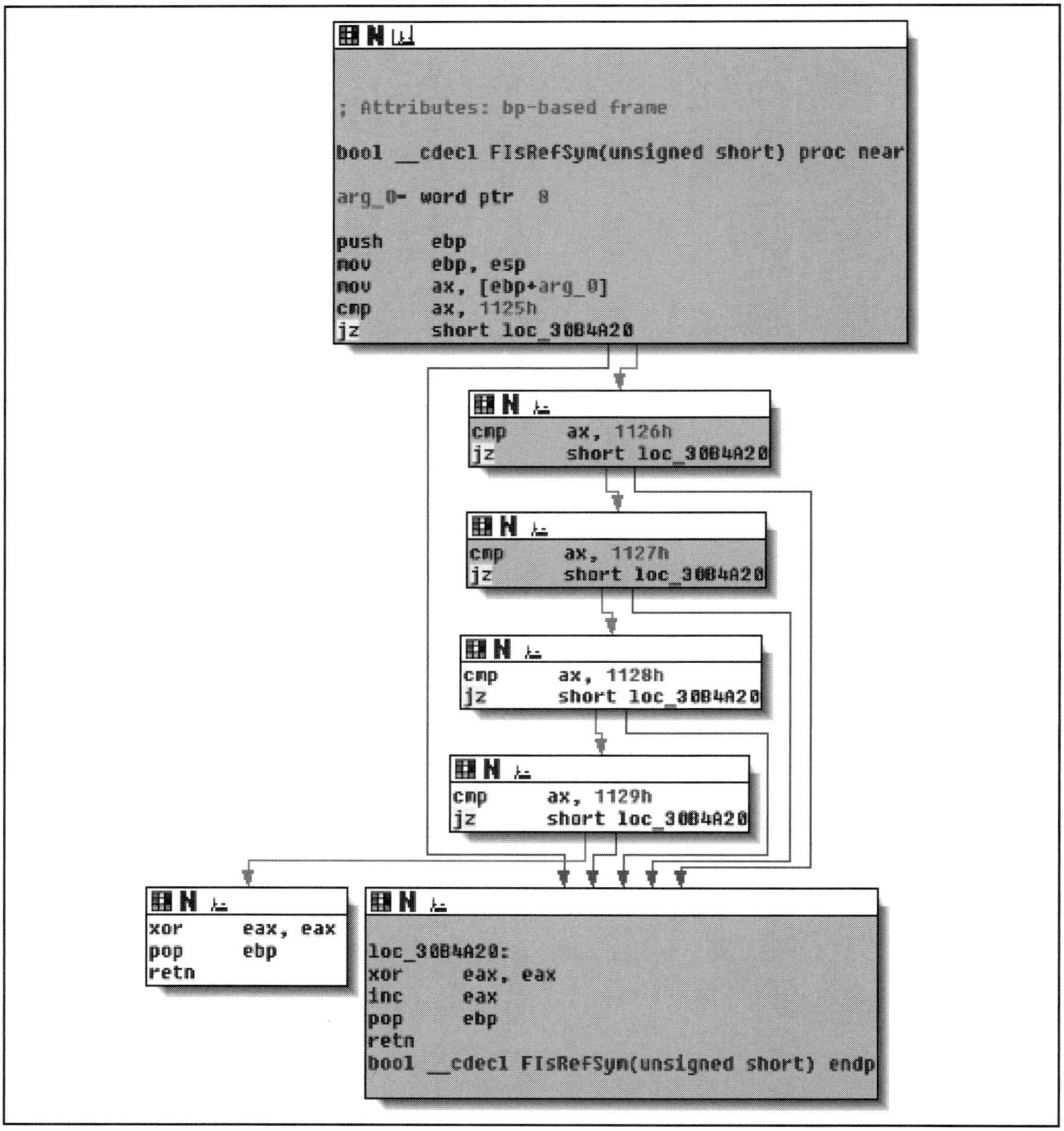

Enter the code from Figure 9.3 into your favorite text editor, preferably a syntax highlighting editor. Save the script with an *.idc* ending. To run use the menu **File | IDC file**.... The script will run, returning control to the user. A new window, **Recent IDC scripts**, will appear Figure 9.5. The left button is *edit* and the right is *execute*. This allows for quick edit cycles.

The code in the script is very similar to some of the snippets in the earlier section. The address range of the current function is determined with the *GetFuncAttr* calls. A loop iterates through the function address range and resets the color using a new IDC functions, *SetColor*.

Figure 9.5 Recent IDC Scripts

While the script is very simple, it is useful for solving an immediate problem. The next section continues this idea while introducing more APIs and concepts.

NOTE

IDA Pro's help file contains documentation for the IDC language. It briefly describes constructs such as statements, expressions, and looping.

The documentation also serves as an API reference for all built in IDC functions.

Writing IDC Scripts

Scripting languages are very popular because of the immediate results they can provide. The user is often trying to solve simple tasks and not developing a full fledged product.

Scripts can and should be written to automate simple tasks. Complete solutions especially in reverse engineering seem to grow organically. Scripting goes hand and hand with this growth. Sometimes we write scripts for specific disassembly projects, while other times the scripts can be used over and over again.

Scripting as well plug-in writing does not have to remove the user from the equation. In the same sense that IDA is an interactive assembler, scripts should be an interactive tool to help the reverse engineer.

Problem solving with IDC

This section is an example of how to use IDC to solve a specific problem. It is not by any means a complete solution, but rather it demonstrates what can be done with very little code and time.

The Problem

C++ uses indirect calls to call many functions. IDA Pro does not create cross references for these functions.

Problem Background

C++ reversing presents some new challenges to the reverse engineer. These challenges like most in reverse engineering can be solved statically or in runtime. Some recent research in static analysis was published by Igor Skochinsky (http://www.openrce.org/articles/full_view/21) (http://www.openrce.org/articles/full_view/23) as a series of articles on the OpenRCE website. Some code examples in the form of IDC scripts are also provided. A paper has also been published by Paul Vincent Sabanal & Mark Vincent Yason from IBM ISS research (https://www.blackhat.com/presentations/bh-dc-07/Sabanal_Yason/Paper/bh-dc-07-Sabanal_Yason-WP.pdf). The paper discusses an internal tool based on IDAPython.

One of the problems with reversing C++ code relates to indirect calls. Code similar to Figure 9.6 is common. A call is made through a register using an offset. It appears that *ecx* is used without being initialized. *Ecx* is passed to the function and represents the *this* pointer. IDA Pro does not know which function is being called and as such does not create a cross reference.

Figure 9.6 An Indirect Call

```
; Attributes: bp-based frame

sub_30CBA25 proc near

arg_0= dword ptr  8
arg_4= dword ptr  0Ch
arg_8= dword ptr  10h

push     ebp
mov      ebp, esp
mov      eax, [ecx]
push     esi
mov      esi, [ebp+arg_8]
lea      edx, [ebp+arg_8]
push     edx
push     [ebp+arg_4]
mov      [ebp+arg_8], esi
push     0
push     [ebp+arg_0]
call     dword ptr [eax+14h]
test     eax, eax
jz       short loc_30CBA4E
```

If we follow execution using a debugger, we can identify the target function. Checking the cross references from the target function will reveal a result similar to Figure 9.7.

Figure 9.7 Xrefs to MSF_HB::ReadStream Before the Script

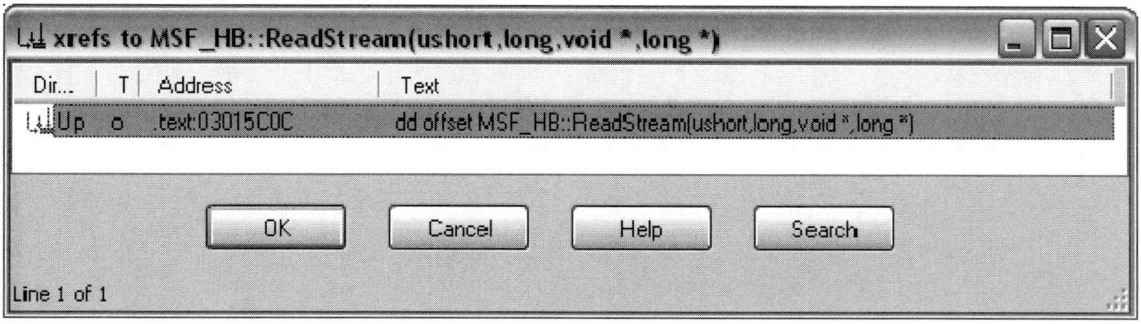

All references are pointers, not function calls. The pointers are part of a *V Table*. A *V Table* is an array of pointers to functions for a particular object.

Figure 9.8 MSF_HB::'vftable'

```
.text:03015BF8 const MSF_HB::'vftable' dd offset
MSF_HB::QueryImplementationVersion(void)
.text:03015BF8; DATA XREF: MSF_HB::MSF_HB(void)+9□o
.text:03015BF8; MSF_HB::~MSF_HB(void)+9□o
.text:03015BFC       dd offset MSF_HB::QueryImplementationVersion(void)
.text:03015C00       dd offset MSF_HB::GetCbPage(void)
.text:03015C04       dd offset MSF_HB::GetCbStream(ushort)
.text:03015C08       dd offset MSF_HB::GetFreeSn(void)
.text:03015C0C       dd offset MSF_HB::ReadStream(ushort,long,void *,long *)
.text:03015C10       dd offset MSF_HB::ReadStream(ushort,void *,long)
.text:03015C14       dd offset MSF_HB::WriteStream(ushort,long,void *,long)
.text:03015C18       dd offset MSF_HB::ReplaceStream(ushort,void *,long)
.text:03015C1C       dd offset MSF_HB::AppendStream(ushort,void *,long)
.text:03015C20       dd offset MSF_HB::TruncateStream(ushort,long)
.text:03015C24       dd offset MSF_HB::DeleteStream(ushort)
.text:03015C28       dd offset MSF_HB::Commit(void)
.text:03015C2C       dd offset MSF_HB::Close(void)
.text:03015C30       dd offset MSF_HB::GetRawBytes(int(*)(void const *,long))
.text:03015C34       dd offset MSF_HB::SnMax(void)
.text:03015C38       dd offset TM::PPdbFrom(void)
.text:03015C3Cdd offset MSF_HB::CloseStream(ulong)
```

When a object method is called, the *VTable* is accessed and then a call is made using an offset to the appropriate function. The code from Figure 9.6 uses this *V Table* making a call to the first *MSF_HB::ReadStream* function.

TIP

Finding *VTables* for Microsoft binaries is simple if using the Determina PDB plug-in by Alexander Sotirov (http://www.determina.com/security.research/utilities/index.html). The *pdb* contains symbols including *VTable* names. Finding all the *VTables* can be done with a text search using the following string:

::‘vftable’ dd

Note that the character before vftable is a back tick, whereas the character following vftable is a single quote.

Proposed solution

In order to find the calling addresses, we could script the debugger and check the stack. Prior to IDA 5.2 IDC functionality regarding the debugger was very limited. The functions did not allow any handling of debugger events, such as a breakpoint. However, there is a workaround using conditional breakpoints.

This script was written long before 5.2 was available and as such does not rely on the new functions. The new functions in 5.2 will be discussed afterwards. Figure 9.9 is the edit breakpoint dialog box. The condition can be any IDC statement including a function call.

Figure 9.9 Setting a Condition to the Handler

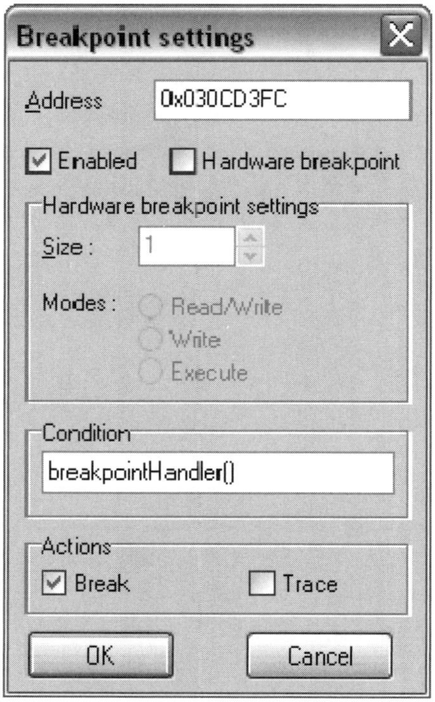

The function will be called when the breakpoint is hit allowing us to run code during the breakpoint. If we don't want to stop the debugger, the function simply returns 0. The function evaluates to false allowing execution to continue. The following code can be used to log the value of *EAX* whenever the breakpoint is hit.

```
static breakpointHandler()
{
    Message("%x bp hit, EAX == 0x%x\n", EIP, EAX);
    return 0; // don't stop on breakpoint
}
```

To check for the caller, we begin by looking at the stack. During a function call, the return address is pushed onto the stack. We need the address to the call instruction which is one instruction before the return address. After the call in Figure 9.6 the return address points to the test instruction rather than the call. This update to the *breakpointHandler* function will log the caller.

```
static breakpointHandler()
{
    auto caller;
    caller = PrevHead(Dword(ESP), (Dword(ESP) - 10));
    Message("%x bp hit, caller == %x\n", EIP, caller);
    return 0; // don't stop on breakpoint
}
```

A new IDC function is introduced, *PrevHead*. The prototype is:

```
long   PrevHead      (long ea, long minea);
```

PrevHead searches for the previous defined instruction or data. The *ea* argument is the start address to begin searching backwards, where *minea* is the lowest address to include in the search. The search in *breakpointHandler* looks up to 10 bytes back. Call instructions through registers are generally only 3 bytes, so the search will find the call. The caller has been determined and a cross reference can be added. The updated *breakpointHandler* adds the cross reference.

```
static breakpointHandler()
{
    auto caller;
    caller = PrevHead(Dword(ESP), (Dword(ESP) - 10));
    AddCodeXref(caller, EIP, XREF_USER | fl_CN);
    return 0; // don't stop on breakpoint
}
```

The *AddCodeXref* adds the cross reference. The prototype is:

```
//      Flow types (combine with XREF_USER!):
#define fl_CF 16        // Call Far
```

```
#define fl_CN 17        // Call Near
#define fl_JF 18        // Jump Far
#define fl_JN 19        // Jump Near
#define fl_F 21         // Ordinary flow
#define XREF_USER 32    // All user-specified xref types
                        // must be combined with this bit
void AddCodeXref(long From,long To,long flowtype);
```

The *Message* function is removed since it slows down the breakpoint handling. Examining other calls in the DLL revealed cross references to be *Call Near*. The *breakpointHandler* function is complete.

Figure 9.10 VTable xref Script

```
#include <idc.idc>
static breakpointHandler()
{
      auto caller;
      caller = PrevHead(Dword(ESP), (Dword(ESP) - 10));
      AddCodeXref(caller,EIP, XREF_USER | fl_CN);
      return 0; // don't stop on breakpoint
}
static setBPs()
{
       auto currAddr;
       auto vStart;
       auto vEnd;
       auto virFunc;
       Message("setBPs() executed\n");
       vStart = SelStart();
       vEnd = SelEnd();
       Message("start = 0x%x\n",vStart);
       Message("end = 0x%x\n",vEnd);
       if ((vStart == BADADDR) || (vEnd == BADADDR))
       {
           Message("No selection made !!\n");
           return;
       }
       if ((vStart - vEnd) %4 != 0)
       {
```

```
        Message("not DWORD aligned\n");
            return;
        }
        for (currAddr = vStart; currAddr < vEnd; currAddr = currAddr + 4)
        {
            virFunc = Dword(currAddr);
            if (GetBptAttr(virFunc, BPTATTR_EA) == -1) // no bpt there yet
            {
                if (!AddBptEx(virFunc, 0, BPT_SOFT))
                {
                    Message("AddBptEx() failed 0x%x\n;", virFunc);
                    return;
                }

                if (!SetBptCnd(virFunc, "breakpointHandler()"))
                {
                    Message("SetBptCnd() failed 0x%x\n;", virFunc);
                    return;
                }
                Message("BP 0x%x set\n", virFunc);
            }
            else
            {
                Message("BP already set 0x%x\n", virFunc);
            }
        }
}
static main()
{
    AddHotkey("Alt-f9", "setBPs");
}
```

Upon running this script, the functions are loaded into memory and *main* is executed. *Main*'s sole purpose is to set a hot key for the *setBPs* function.

A *VTable* similar to Figure 9.8 is selected and then the hotkey is pressed. The *setBPs* function is called by the hotkey. This functions purpose is to set breakpoints on the targets found in the *VTable*. Only one breakpoint can be added per address, thus the function checks for the presence of a breakpoint initially. If no breakpoint exists a new software breakpoint is added. The breakpoint is then made conditional, with the condition being the *breakpointHandler* function. In this case we choose to return 0 and not stop at the breakpoint.

The result after using the script and running the debugger is shown in Figure 9.11, which is quite an improvement over the original in Figure 9.7.

Figure 9.11 Xrefs to MSF_HB::ReadStream after the Script

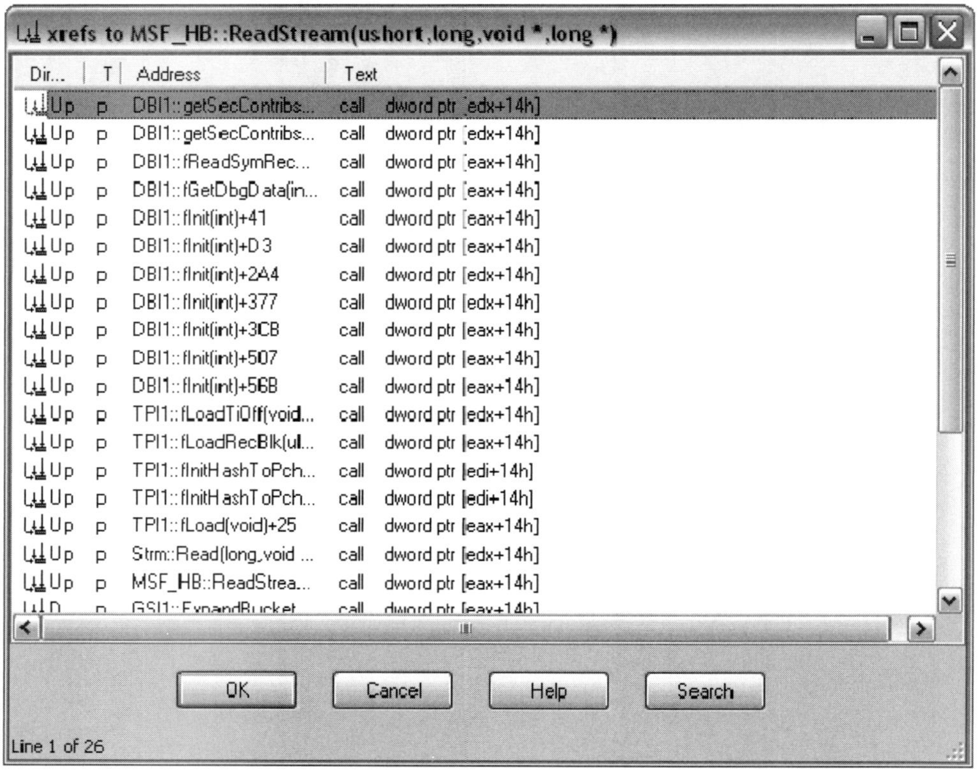

Possible Improvements

If the breakpoint is set on a commonly called function, performance can be degraded. A global variable can be used as a counter and the breakpoint could be removed if the counter reaches a preset limit. Comments can be added to the calling instruction containing the address of the targets. This would allow the user to double click the address to the target function.

After acquiring the cross reference data, we could analyze it to determine information of the classes they represent. Method visibility in C++ can be public, protected, or private.

Public	The target function has at least one call from a function not found in any VTable.
Private	The target function is only called from other functions within its own VTable.
Protected	The target function is only called from functions located in VTables.

If a function is located in more than one *VTable* at the same offset, it is likely that an inheritance relationship exists between the classes.

The cross reference information could be combined with static analysis to reconstruct object models. A graphing tool could be used to represent the models, perhaps in UML.

New IDC Debugger Functionality

New IDC functions are added to IDA Pro releases, while it is rare for functions to be deprecated. The new functions reflect the new features added to the release.

IDA 5.2 added 53 functions. The most important additions relate to the scripting of the debugger http://www.hex-rays.com/idapro/scriptable.htm. The debugger is now fully scriptable from IDC. We can control every aspect of the debugger. This includes acting on debugger events, attaching to processes, and tracing.

A scriptable debugger opens up many possibilities. Unpacking of binaries for a known packer is commonly scripted using OllyScript for Ollydbg or in Python for Immunity Debugger. Runtime analysis can be fed back into the static analysis.

The debugger from a plug-in context usually requires a callback to handle events. The IDC interface allows for what amounts to a blocking call waiting on events. The function used is *GetDebuggerEvent* and its prototype is:

```
long GetDebuggerEvent(long wfne, long timeout);
```

The timeout can be set to −1, which is interpreted as infinite. There wfne flags are the following:

```
WFNE_ANY      return the first event
WFNE_SUSP     wait until the process gets suspended
WFNE_SILENT set: be slient, clear:display modal boxes if necessary
WFNE_CONT     continue from the suspended state
```

Most often we want to wait till for a suspended state caused by a breakpoint. The possible return values are the following:

```
// debugger event codes
NOTASK          process does not exist
DBG_ERROR       error (e.g. network problems)
DBG_TIMEOUT     timeout
PROCESS_START   New process started
PROCESS_EXIT    Process stopped
THREAD_START    New thread started
THREAD_EXIT     Thread stopped
BREAKPOINT      Breakpoint reached
STEP            One instruction executed
EXCEPTION       Exception
LIBRARY_LOAD    New library loaded
```

```
LIBRARY_UNLOAD      Library unloaded
INFORMATION         User-defined information
SYSCALL             Syscall (not used yet)
WINMESSAGE          Window message (not used yet)
PROCESS_ATTACH      Attached to running process
PROCESS_DETACH      Detached from process
```

This addition to IDC is very welcome as it provides easy scripting use of the debugger. A plug-in that works with the debugger is presented later in this chapter.

Useful IDC Functions

This section contains a sampling of IDC functions you are likely to see in other scripts and while writing new scripts. The functions are grouped into similar categories and include a short description of possible usage.

Reading and Writing Memory

Reading memory is accomplished through three functions. The functions come in variants based on the read size. The functions are *Byte*, *Word*, and *Dword*.

```
long   Byte (long ea);       // get a byte at ea
long   Word (long ea);       // get a word (2 bytes) at ea
long   Dword (long ea);      // get a double-word (4 bytes) at ea
```

The functions return a −1 on failure. In order to differentiate between a failure and a value of −1, the macro *hasValue* should be called. Its prototype is:

```
#define hasValue(F)     ((F & FF_IVL) != 0)       // any defined value?
```

The macro will return a 0 if the value is not defined.

Writing to memory is accomplished by the *Patch* family of functions. The functions are used for writing within the static analysis within the IDB as well as virtual memory when under a debugger. In fact the *Patch* functions are the only way to modify code during execution within the debugger. The debugger only allows modification of registers and sections from the GUI. Modification to code or data segments requires these IDC functions or a plug-in.

The function come in three variant based on the write size. The functions are *PatchByte*, *PatchWord*, and *PatchDword*.

```
void     PatchByte      (long ea,long value);     // change a byte
void     PatchWord      long ea,long value);      // change a word (2 bytes)
void     PatchDword     (long ea,long value);     // change a dword (4 bytes)
```

Cross References

There are different types of cross references, both for data and code.

Code Xrefs

Code cross references are defined by their flowtypes. The following is a list of code flowtypes:

```
//      Flow types (combine with XREF_USER!):
#define fl_CF   16       // Call Far
#define fl_CN   17       // Call Near
#define fl_JF   18       // Jump Far
#define fl_JN   19       // Jump Near
#define fl_F    21       // Ordinary flow
#define XREF_USER 32   // All user-specified xref types
                      // must be combined with this bit
```

All user created referenced should be combined with *XREF_USER*. We used *fl_CN*, call near flowtype in the script. There are also near and far jump flowtypes. The ordinary flowtype is used between consecutive instructions.

IDC functions for adding and deleting code cross references:

```
void   AddCodeXref(long From,long To,long flowtype);
long   DelCodeXref(long From,long To,int undef);
```

The *undef* argument undefines the *To* address if this is the last reference to it.

There are two sets of IDC functions to iterate through references. The difference is in regards to recognizing ordinary flows as cross references. The first set will return the ordinary flow first.

```
long   Rfirst   (long From);               // Get first code xref from 'From'
long   Rnext    (long From,long current);// Get next code xref from
long   RfirstB  (long To);                 // Get first code xref to 'To'
long   RnextB   (long To,long current); // Get next code xref to 'To'
```

The functions consist of a *first* and a *next*. Both functions are generally used in a loop to iterate through cross references. The following demonstrates these functions:

```
auto xfAddr, origAddr;
origAddr = ScreenEA();
xfAddr = RfirstB(origAddr);
while (xfAddr != BADADDR)
{
    Message("%x to %x, type == %d\n", xfAddr, origAddr, XrefType());
    xfAddr = RnextB(origAddr, xfAddr);
}
```

The code iterates through all the cross references for the address the cursor is on. It also introduces a new IDC function *XrefType*. The prototype is:

```
long XrefType(void); // returns type of the last xref
                     // obtained by [RD]first/next[B0]
                     // functions. Return values
                     // are fl_... or dr_...
```

XrefType return the type of the last cross reference accessed. This function also works on data cross references which will be discussed shortly. The second set of code cross reference functions mirrors the first set.

```
long   Rfirst0 (long From);
long   Rnext0 (long From,long current);
long   RfirstB0(long To);
long   RnextB0 (long To,long current);
```

These functions do not return ordinary flow cross references.

Data Xrefs

The following are valid data types:

```
//      Data reference types (combine with XREF_USER!):
#define dr_O     1                  // Offset
#define dr_W     2                  // Write
#define dr_R     3                  // Read
#define dr_T     4                  // Text (names in manual operands)
#define dr_I     5                  // Informational

#define XREF_USER 32            // All user-specified xref types
                                // must be combined with this bit
```

Same as code *xrefs*, user made data *xrefs* should be combined with *XREF_USER*. Data *xrefs* have only one set of functions associated with them.

```
long     Dfirst      (long From);     // Get first data xref from 'From'
long     Dnext       (long From,long current);
long     DfirstB     (long To);       // Get first data xref to 'To'
long     DnextB      (long To,long current);
```

Data Representation

Data representation functions create structures, functions, data, and define code among other things. They are the IDC function equivalent to many manual tasks done during disassembly. The following is a sampling of these functions.

```
success    MakeArray(long ea,long nitems);
success    MakeByte(long ea);
long       MakeCode(long ea);
success    MakeData(long ea, long flags, long size, long tid);
success MakeDword(long ea);
```

```
success MakeFunction(long start,long end);
success MakeStr(long ea,long endea);
success MakeStructEx(long ea,long size, string strname);
```

Comments

Comments are a key to successful reverse engineering. Comments along with proper naming are the notes that bind to the binaries we analyze. There are IDC functions to set and read comments.

```
// repeatable, 0 = standard, 1 = repeatable
string CommentEx(long ea, long repeatable);
success MakeComm(long ea,string comment);
success MakeRptCmt(long ea,string comment);
long SetBmaskCmt(long enum_id,long bmask,string cmt,long repeatable);
success SetConstCmt(long const_id,string cmt,long repeatable);
success SetEnumCmt(long enum_id,string cmt,long repeatable);
void SetFunctionCmt(long ea, string cmt, long repeatable);
long SetMemberComment(long id,long member_offset,string comment,long repeatable);
long SetStrucComment(long id,string comment,long repeatable);

long GetBmaskCmt(long enum_id,long bmask,long repeatable);
string GetConstCmt(long const_id,long repeatable);
string GetEnumCmt(long enum_id,long repeatable);
string GetFunctionCmt(long ea, long repeatable);
string GetMarkComment(long slot);
string GetStrucComment(long id,long repeatable);
```

Code Traversal

IDA has different types of containers for code and data. Some of the containers include segments, functions and instruction or data heads. Iterating through different containers and areas is very common in scripts.

Some common iterating functions are:

```
long NextAddr(long ea);
long NextFunction(long ea);
long NextHead(long ea, long maxea);
long NextNotTail(long ea);
long NextSeg(long ea);

long PrevAddr(long ea);
long PrevFunction(long ea)
long PrevHead(long ea, long minea);
long PrevNotTail(long ea);
```

The following code snippet demonstrates some of the iteration functions.

```
auto currAddr, func, endSeg,funcName, counter;

currAddr = ScreenEA();
func = SegStart(currAddr);
endSeg = SegEnd(currAddr);

counter = 0;
while (func != BADADDR && func < endSeg)
{
         funcName = GetFunctionName(func);
         if (funcName != " ")
         {
                  Message("%x: %s\n", func, funcName);
                  counter++;
                  }
                  func = NextFunction(func);
}

Message ("%d functions in segment: %s\n", counter, SegName(currAddr));
```

The script iterates through all the functions belonging to the current segment. The script uses the *GetFunctionName* call to test if an address is in a function. This call returns an empty string if the address is not part of a function. Alternatively, *GetFunctionFlags* could have been used. The script prints a list of function addresses along with names for all the functions in the segment. The total number of functions is printed upon completion.

Input and Output

Thus far the only I/O used has really been the *Message* function. There are various input IDC functions for different types of data as well as for making selections.

```
string    AskStr(string defval,string prompt);
string    AskFile(bool forsave,string mask,string prompt);
long      AskAddr(long defval,string prompt);
long      AskLong(long defval,string prompt);
long      AskSeg(long defval,string prompt);
string    AskIdent(string defval,string prompt);
long      AskYN(long defval,string prompt);
```

The preceding functions retrieve input from the user. The following code snippet demonstrates the *AskYN IDC* function.

```
auto answer;
answer = AskYN(1, "hello");
```

```
if (answer == 1)
   Message("YES\n");
else if (answer == 0)
   Message("NO\n");
else
   Message("CANCEL\n");
```

There are IDC functions for file I/O as well. These file I/O functions are very similar to their C counterparts.

```
long fopen(string file,string mode);
long fseek(long handle,long offset,long origin);
void fclose(long handle);

long fgetc(long handle);
long fprintf(long handle,string format,…);
long fputc(long byte,long handle);
long ftell(long handle);

long writelong(long handle,long dword,long mostfirst);
long writeshort(long handle,long word,long mostfirst);
long writestr(long handle,string str);

long readlong(long handle,long mostfirst);
long readshort(long handle,long mostfirst);
string readstr(long handle);
```

Basics of IDA Plug-ins

IDA Pro can be extended through modules. There are various types of modules that can be developed for IDA. Plug-ins are one of the types of modules that can be used to extend IDA. Sometimes the term plug-in is used incorrectly to cover all extendable modules.

There are different types of modules available in IDA. The module type is dependent on the functionality needed. The categories are:

- Plug-in
- Loaders
- Processor
- Debuggers

Module/Plug-in Resources

The SDK has many modules/plug-ins with full source code.

- Hex-Rays provides an SDK to registered customers. The SDK is included on the CD when IDA is purchased and is also available on the Hex-Rays website (http://www.hex-rays.com/idapro/idadown.htm). The SDK is not officially supported by Hex-Rays, although the option is available.

- The Hex-Rays bulletin board provides help with plug-in issues both by other users as well as Ilfak (http://www.hex-rays.com/forum/).
- There is not much information available for plug-in development. An excellent tutorial was written by Steve Micallef entitled "IDA PLUG-IN WRITING IN C/C++" (http://binarypool.com/idapluginwriting/).
- Hex-Rays includes various plug-ins with source in the SDK. One of the plug-ins, a universal unpacker, is described in a Hex-Rays article (http://www.hex-rays.com/idapro/unpack_pe/unpacking.pdf).
- Ilfak has also provided various plug-ins with full source available at his blog (http://hexblog.com).
- OpenRCE is a valuable resource for many thing reverse engineering. A portion of the download site is dedicated to IDA Pro plug-ins (http://www.openrce.org/downloads/browse/IDA_Plugins).

Plug-ins are written in C++ and support a variety of compilers and development environments. Import libraries are provided for the following:

- Visual C++ (32 and 64 bit)
- Borland C++ Builder (32 and 64 bit)
- GCC C++ Compiler
- Windows (32 and 64 bit)
- Linux (32 and 64 bit)
- Mac OSX (32 and 64 bit)

This chapter will focus on 32 bit Windows plug-ins using the Microsoft Visual Studio 2005/2008 compilers.

NOTE

Instructions for other development environments are located in the root directory of the SDK.

install_cb.txt contains CBuilder setup instructions.
install_mac.txt contains OS X GCC setup instructions.
install_linux.txt contains Linux GCC setup instructions.

Processor modules add support for different CPUs and architectures. Processor modules are located in the *procs* directory. These modules interpret opcodes and generate the disassembly we see in IDA.

Processor modules use the following file extensions:

- **w32** windows
- **w64** windows 64
- **ilx** Linux
- **ilx64** Linux 64
- **imc** OS X
- **imc64** OS X 64

Loaders operate similar to operating system loaders. A loader parses executable files, creates segments, and determines what segments are code or data. IDA includes loaders for various executable files including PE (Portable Executables) and ELF (Executable and Linking Format).

Loader modules use the following file extensions:

- **ldw** windows
- **l64** windows 64
- **llx** Linux
- **llx64** linux 64
- **lmc** OS X
- **lmc64** OS X 64

Debugger modules are complete debuggers are interoperate with IDA. These should not be confused with standard plug-ins that work with the built-in debugger modules. Documentation for debugger modules consist of source code located in the SDK under *\plugins\debugger.*

Standard plug-ins encompass everything else not covered by either the processor or loader modules. They are commonly referred to just as plug-ins. These plug-ins operate on the disassembly. This is the most common type of plug-in and as such will be documented in this chapter.

Standard plug-ins use the following file extensions:

- **plw** windows
- **p64** windows 64
- **plx** Linux
- **plx64** Linux 64

- **pmc** OS X
- **pmc64** macosx64

Introducing the IDA Pro SDK

Hex-Rays froze the SDK starting with version 4.9. What does this mean to us? Previously the SDK changed considerably between different versions. Plug-ins needed to be compiled for each SDK as they were not binary compatible. We no longer have to worry about big changes between SDKs, besides the addition of new functionality.

The SDK is on the IDA Pro CD, or is available for download from Hex-Rays (http://www.hex-rays.com/idapro/idadown.htm).

The SDK is a zip file, the latest being idasdk52. I have directory for different SDKs, so I extract the archive to \SDK\idasdk52.

WARNING

The SDK may contain a bug that will prevent proper compilation. The bug is in the intel.hpp located in \include. One of the #include listing is wrong. Change

#include "../idaidp.hpp"

To

#include "../module/idaidp.hpp"

The bug is still present in the latest version 5.2.

SDK Layout

The SDK contains various directories. The directories contain, include files, import libraries, tools, and source code. The following is an overview of the more important directories:

include	All the header files for the SDK.
ldr	Source code to several loaders.
libbor.w32	Borland for 32 bit Windows plugins.
libbor.w64	Borland for 64 bit Windows plugins.
libgcc.w32	GCC for 32 bit Windows plugins.
libgcc.w64	GCC for 32 bit Windows plugins.
libgcc32.lnx	GCC for 32 bit Linux plugins.
libgcc32.mac	GCC for 32 bit OS X plugins.
libgcc64.lnx	GCC for 64 bit Linux plugins.
libgcc64.mac	GCC for 64 bit OS X plugins.
libvc.w32	Visual Studio for 32 bit Windows plugins.

libvc.w64	Visual Studio for 64 bit Windows plugins.
module	Source code to several processor modules.
plugins	Source code to sample and real plugins.

Plug-in Syntax

Plug-ins are loadable libraries, DLL or otherwise, that IDA Pro loads when needed. The plug-in has to have a certain structure exported. The structure type depends on the plug-in type. This section will cover standard plug-ins as they are the most common type. From now on plug-ins will refer to standard plug-ins. Any specifics relating to loaders or processor modules will be noted.

IDA plug-ins are written in C++ and export a plug-in structure, *PLUGIN_t*.

```
plugin_t PLUGIN =
{
  IDP_INTERFACE_VERSION,
  plugin_flags,       // plugin flags
  init,               // initialize
  term,               // terminate. this pointer may be NULL.
  run,                // invoke plugin
  comment,            // plugin comment
  help,               // multiline help about the plugin
  wanted_name,        // the preferred short name of the plugin
  wanted_hotkey       // the preferred hotkey to run the plugin
};
```

The structure contains constants, function pointers, and character string pointers.

IDP_INTERFACE_VERSION is a constant that will be defined by included SDK files. Previous to the 4.9 SDK freeze, this value would be incremented for every new release. Since 4.9 this value has remained constant.

The *plugin_flags* define how the plug-in operates with IDA. The different flags are described in *loader.hpp*. This field is usually set to 0 or set to *PLUGIN_UNL* when debugging a plug-in.

The following three items, *init, term*, and *run* are function pointers.

The *init* function is the executed when the plug-in is loaded. Its main purpose is to determine if the plug-in is applicable to the current database. Plug-ins can be specific to processors or file formats. Additionally this function could setup the environment for the plug-in once run is executed.

The *init* function needs to return one of the following:

- **PLUGIN_SKIP** This notifies IDA to not load the plug-in. A plug-in usually returns this value, when the architecture or file format isn't appropriate. For example:

```
if (inf.filetype != f_PE)
   return PLUGIN_SKIP; // not a PE file
```

- **PLUGIN_OK** This notifies IDA that the plug-in is appropriate and IDA will load the plug-in upon first use.
- **PLUGIN_KEEP** This notifies IDA that the plug-in is appropriate and to leave the plug-in in memory.

The term function is executed when the IDA is being terminated. This function can be used to clean up resources used during the plug-ins lifetime. Many plug-ins set this pointer to NULL.

The run function is executed by running the plug-in. This function accepts arguments. The arguments are defined within the plugin.cfg file located in the plugin directory. Many plug-ins use the run function to do all the necessary work. Other plug-ins use the run function to set up callbacks. Debugging functionality in the SDK is handled by callbacks.

- **Comment** is pointer to a short character string description for the plug-in.
- **Help** is also a pointer to a character string. However unlike the Comment string, Help is usually a multiline description of the plug-in.
- **Wanted_name** is the name that is displayed in the plug-in list accessible from (File | Edit | Plugins).
- **Wanted hotkey** sets up a hotkey to run the plug-in. This hotkey can be overridden via plugins.cfg.

Currently the comment and help fields are not used by IDA, but this may change in the future.

Setting up the Development Environment

This section covers setting up the development environment under Visual Studio 2005 and 2008. Build instructions for other platforms are available in the base directory of the SDK.

Setting up development environments can be tedious. The easiest way to start writing plug-ins is to use the IDA Pro Plug-in Wizard. The wizard is compatible with Visual Studio 2005 and 2008. All appropriate compiler and linker options will be configured by the wizard.

The IDA Pro Plug-in Wizard is available from http://ringzero.net/re. The wizard is compatible with:

- Visual Studio 2008
- Visual Studio 2005
- Visual C++ 2008 Express Edition
- Visual C++ 2005 Express Edition

Tools & Traps…

Building Plug-ins under Linux

Setting up a proper plug-in build environment under Linux can be complicated. The following *makefile* can be used to build plug-ins. Note that command lines must begin with a tab. The lines following the *all*, *install*, and *clean* labels are command lines.

```
# Makefile for IDA Pro Plugins under Linux
# Updated version of makefile from Steve Micallef's
# IDA Plugin Writing Tutorial
# http://www.binarypool.com/idapluginwriting/

# Set your plugin name here. PLUGINNAME.plx
PLUGINNAME=myplugin

# Set your IDA install directory
IDABASEDIR=/usr/local/idaadv

# Set your IDA SDK directory
SDKBASEDIR=/usr/local/idaadv/sdk

# Compiles all cpp files in current dir
SRC=$(wildcard *.cpp)
OBJS=$(SRC:.cpp=.o)
CC=g++
LD=g++
CFLAGS=-D__IDP__ -D__PLUGIN__ -c -D__LINUX__ \
    -I$(SDKBASEDIR)/include $(SRC)
LDFLAGS=--shared $(OBJS) -L$(IDABASEDIR) -lida -no-undefined \
    -Wl,-version-script=$(SDKBASEDIR)/plugins/plugin.script

all:
	$(CC) $(CFLAGS)
	$(LD) $(LDFLAGS) -o $(PLUGINNAME).plx

install:
	cp $(PLUGINNAME).plx $(IDABASEDIR)/plugins
```

Continued

```
clean:
  -rm -f *.plx *.o core

rebuild: clean all
```

Simple Plug-in Examples

Now that we have setup a development environment, we can move on to writing some plug-ins. We will first build a simple "hello world" plug-in. This will allow us to test our environment and verify that IDA is properly loading and executing our plug-in. The *find memcpy* plug-in will demonstrate some of the IDA API including instruction decoding and some UI code.

The Hello World Plug-in

Start Visual Studio and select the IDA Pro Plugin Wizard. Enter the project name and click **OK**. Figure 9.12 shows the selection within Visual C++ 2008 Express Edition.

Figure 9.12 Selecting the IDA Pro Plug-in Wizard

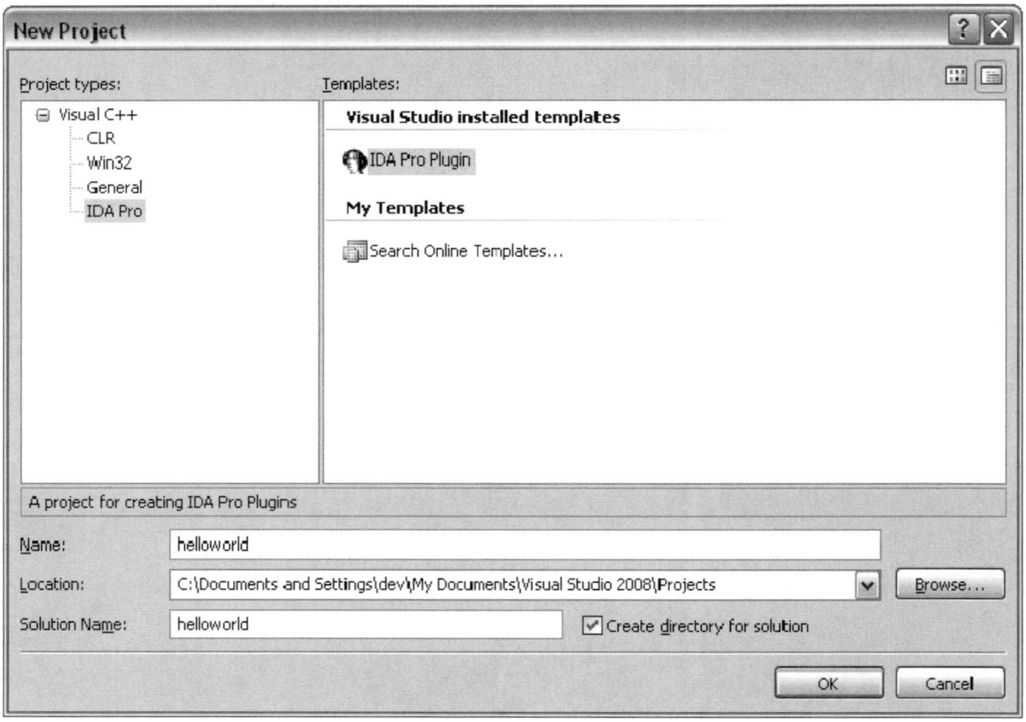

The wizard is shown in Figure 9.13. The plug-in type will default to plug-in, which is what we are building. The *Name of Author* field is optional, but will appear in the header comments if present.

The SDK Path is required to build the plug-in. Use the button to bring up the folder browse dialog. Be sure to select the base on the SDK directory.

The final item is optional but very useful. A post build event is created in the project properties. The event copies the plug-in to the appropriate IDA Pro directory. Use the button to bring up the folder browse dialog. Be sure to select the base on the IDA Pro install directory.

The paths only need to be filled in once as the wizard saves the options.

Figure 9.13 IDA Pro Plugin Wizard Dialog

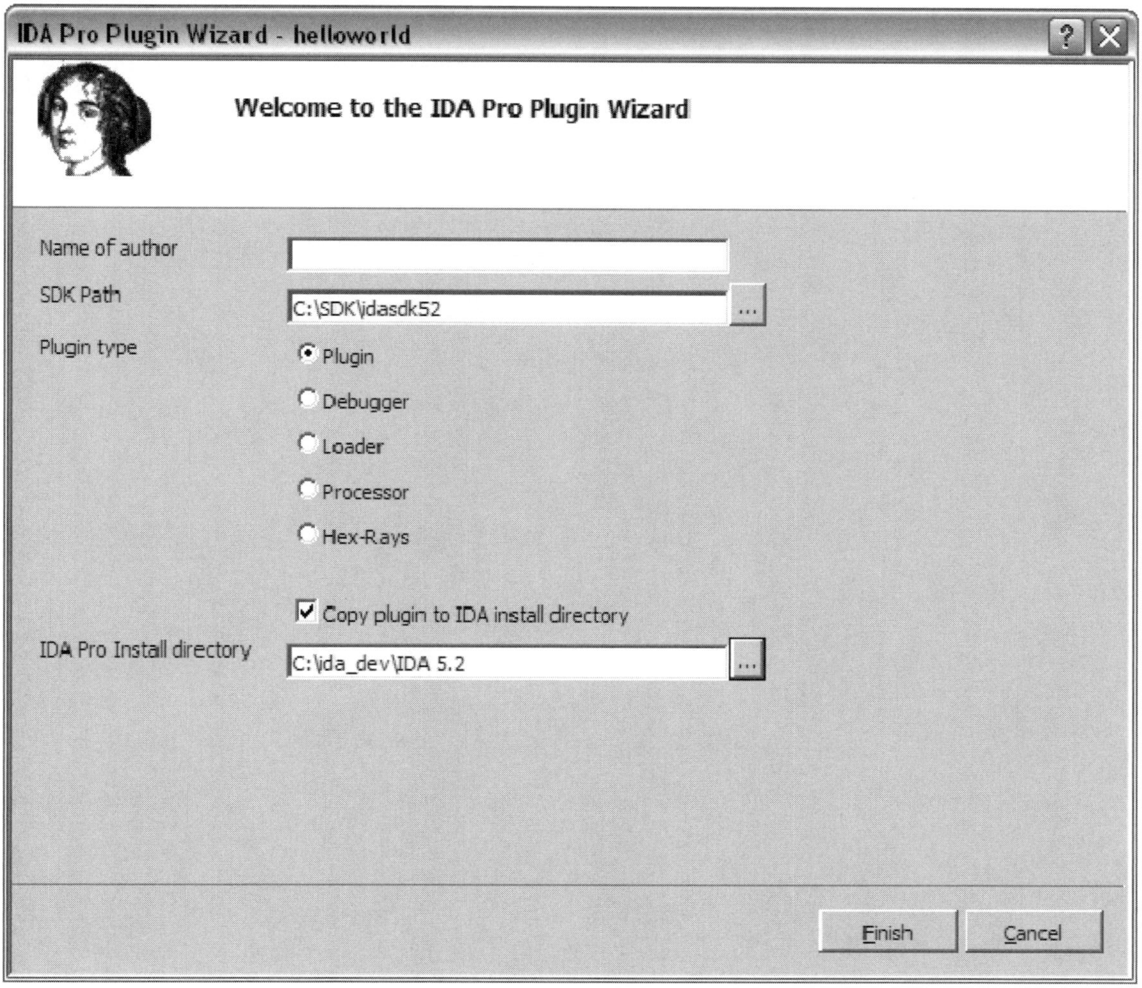

Click Finish and the wizard will complete preparing the project. The IDA Pro Plug-in wizard will have a sample template. The following code can be copied over the template. The key combination to start a build varies on configuration, but by default it is **CRTL+SHIFT+B**.

```
#include <ida.hpp>
#include <idp.hpp>
#include <loader.hpp>

// Determine if the plugin is suitable. Return:
// PLUGIN_SKIP - plugin not suitable, wont be used
// PLUGIN_KEEP       - plugin is suitable. keep in memory
// PLUGIN_OK  - plugin is suitable, load when used
int init(void)
{
    return PLUGIN_OK;
}

// plugin termination function. Unhook any notification points.
void term(void)
{
    return;
}

// This function is called when the plugin is executed.
// The arg is configured in the plugin.cfg file.
void run(int arg)
{
    msg("Hello world! my address is %a\n", get_screen_ea());
    return;
}

char comment[] = "hello world";
char help[] = "hello world";

// Name of plugin in ( Edit | Plugins )
// An entry in plugins.cfg can overide this field.
char wanted_name[] = "hello world";

// Plugin's hotkey
// An entry in plugins.cfg can overide this field.
char wanted_hotkey[] = "";
```

```
// PLUGIN DESCRIPTION BLOCK
plugin_t PLUGIN =
{
IDP_INTERFACE_VERSION,
PLUGIN_UNL,        // plugin flags
init,              // initialize
term,              // terminate. this pointer may be NULL.
run,               // invoke plugin
comment,           // comment about the plugin
help,              // multiline help about the plugin
wanted_name,       // the preferred short name of the plugin
wanted_hotkey      // the preferred hotkey to run the plugin
};
```

The previous code is the equivalent of a hello world program. It implements the key functions outlined in the *plugin_t* structure. The plug-in uses a flag, *PLUGIN_UNL*. This flag is often used for debugging plug-ins. It is defined as:

```
#define PLUGIN_UNL 0x0008     // Unload the plugin immediately after
                              // calling 'run'.
                              // This flag may be set anytime.
                              // The kernel checks it after each
                              // call to 'run'
                              // The main purpose of this flag is to ease
                              // the debugging of new plugins.
```

The plug-in will be unload after run is executed. This allows us to make changes, recompile, and copy the plug-in to the plug-in directory. If the flag is not set, IDA needs to be restarted as it retains an open file handle to the plug-in. A workaround will be presented shortly. Note that unloading only occurs after executing run. If the init function returns *PLUGIN_KEEP*, the plug-in remains in memory and will not be unloaded until the run function is executed. However if the init function returned *PLUGIN_OK*, plug-in is only loaded upon first use.

WARNING

If the init function sets callbacks it must return *PLUGIN_KEEP*. Otherwise, the memory addresses used may become invalid, as the plug-in may load at a different address.

The plug-in's run function outputs a message containing "hello world" and the current address. The IDA API call used is *get_screen_ea*. This function is equivalent to the IDC function used earlier in this chapter, *ScreenEA*. The IDA API contains equivalents to many IDC functions as well as functionality not available from IDC. The hello world plug-in verifies that we have a working development environment.

The find memcpy Plug-in

With a working development environment we can move on to more useful plug-ins, while introducing some new IDA API calls. This plug-in searches for inline *memcpys* (See Figure 9.14). Compilers often inline library calls as an optimization. *Memcpy* is commonly inlined along with string functions such as *strlen* and *strcpy*.

The assembly in Figure 9.14 uses *movsd* and *movsb* instructions to copy data. The movs instructions operate on the *edi* and *esi* register. *Esi* holds the source address, while *edi* points to the destination. *Movsd* copies a *dword* (four bytes) and *movsb* copies a single byte.

rep is a prefix that *repeats* the *movs* instructions. Every time *movs* instruction executes *ecx* is decremented. The movs instructions stop once *ecx* reaches zero. (See Figure 9.15).

Figure 9.14 Inline memcpy

```
                    ; memcpy (edi, esi, eax)
.text:00418ECC    mov ecx, eax    ; copy eax into ecx
.text:00418ECE    shr ecx, 2      ; shift ecx right by 2 (divide by 4)
.text:00418ECE                    ; ecx = number of dwords to copy
.text:00418ED1    rep movsd       ; copy dwords from esi to edi
.text:00418ED3    mov ecx, eax    ; copy eax into ecx
.text:00418ED5    and ecx, 3      ; and ecx by 3
.text:00418ED5                    ; ecx = number of bytes to copy
.text:00418ED5                    ; (remaining bytes)
.text:00418ED8    rep movsb       ; copy bytes from esi to edi
```

Figure 9.15 rep movsd flowchart

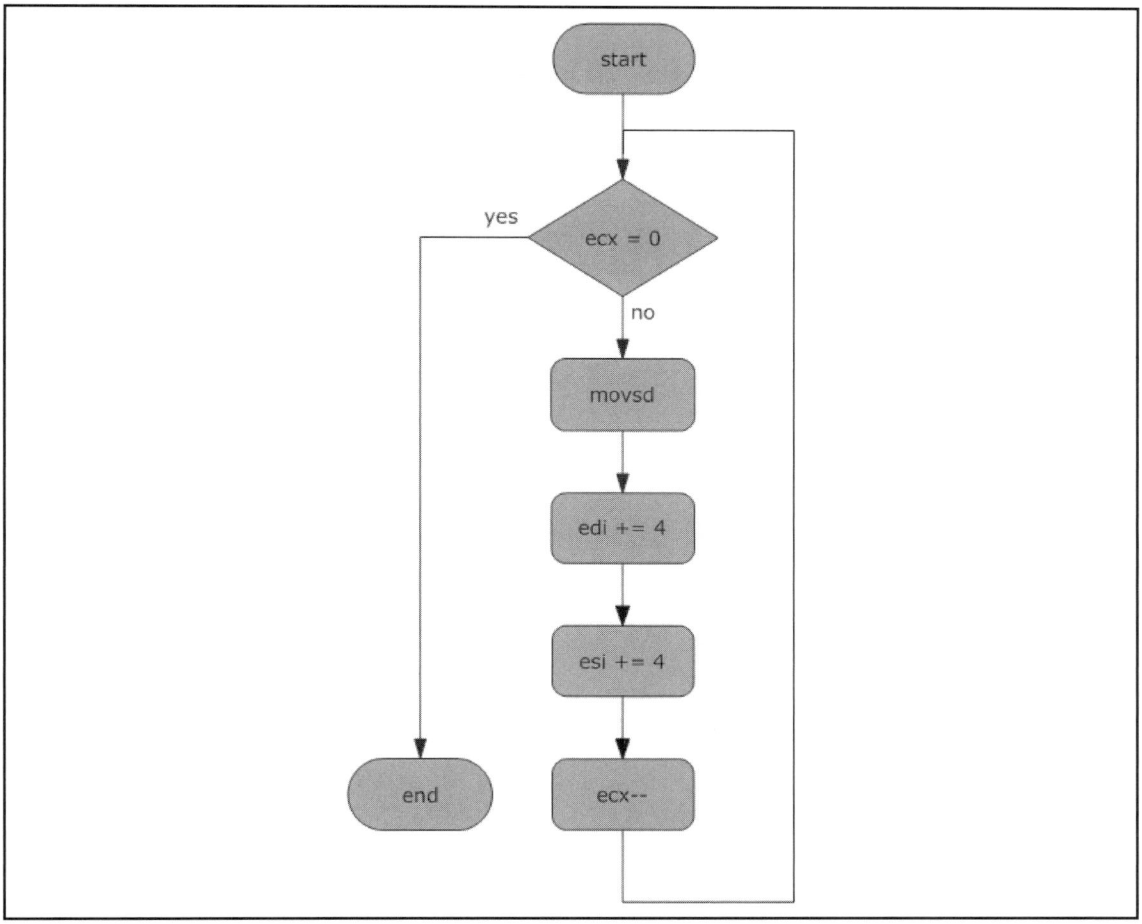

Eax contains the number of bytes to copy. The *shr* (shift right) instruction divides *ecx* by four, calculating the number of dwords to copy. The *rep movsd* instruction copies *ecx* dwords from *esi* to *edi*. After copying the *dwords*, there can be zero to three bytes left to copy. The *and* instruction calculates the remaining bytes which *rep movsb* copies.

The plug-in in Figure 9.16 illustrates a method for finding these types of code constructs.

Figure 9.16 Find memcpy Plug-in

```
/*****************************************************************
* Find memcpy() IDA Pro plugin
*
* Copyright (c) 2008 Luis Miras
* Licensed under the BSD License
*
```

```
* Requirements: The plugin requires x86 processor.
*
* Description: The plugin searches for rep movsd/rep movsb
*              pairs identidying them as memcpy()
*              Single rep movsd and rep movsb instructions
*              are also recorded
*
* Data structures: a netnode is the main data structure.
*                  movsobj_t represents the either pairs
*                  or single instructions.
*
* netnodes are implemented internally as B-trees.
* IDA uses netnodes extensively for its own storage.
* netnodes are defined in netnode.hpp.
*
* netnodes in the plugin: calls - holds all indirect calls
*                         vtable - holds all vtables
*
* netnodes have various internal data structures.
* The plugin uses 2 types of arrays:
*    altval - a sparce array of 32 bit values, initially set to 0.
*    supval - an array of variable sized objects (MAXSPECSIZE)
*
* The plugin holds base addresses in altval and movsobj_t objects
* in supval
*****************************************************************/

#include <ida.hpp>
#include <idp.hpp>
#include <loader.hpp>
#include <allins.hpp>
#include <intel.hpp>

#define NODE_COUNT -1

struct movsObj {
   ea_t movsDW; // addr of rep movsd. BADADDR if none
   ea_t movsBT; // addr of rep movsd. BADADDR if none
};

typedef movsObj movsobj_t;
```

```
static const char* header[] = {"Address", "Type", "Movsd/b distance"};
static const int widths[] = { 16, 25, 25};
char window_title[] = "Inline memcpy" ;

/*******************************************************************
* Function: processMemcpy
*
* This function determines the types of memcpy based on the movsobj_t
* and calculates distance between rep movsd and rep movsb
*******************************************************************/
char* processMemcpy(movsobj_t* my_movs, ea_t* movs_distance)
{
      if (my_movs->movsDW == BADADDR)
      {
          *movs_distance = BADADDR;
          return "memcpy movsb only";
      }
      else if (my_movs-> movsBT == BADADDR)
      {
          *movs_distance = BADADDR;
          return "memcpy movsd only";
      }
      else
      {
          *movs_distance = my_movs-> movsBT - my_movs->movsDW;
          return "memcpy()";
      }
}
/********************************************************************
* Function: description
*
* This is a standard callback in the choose2() SDK call. This function
* fills in all column content for a specific line. Headers names are
* set during the first call to this function, when n == 0.
* arrptr is a char* array to the column content for a line.
*                 arrptr[number of columns]
*
* description creates 3 columns based on the header array
********************************************************************/
```

```
void idaapi description(void *obj,ulong n,char * const *arrptr)
{
      netnode *node = (netnode *)obj;
      movsobj_t my_movs;
      char* outstring = NULL;
      ea_t movs_distance;

      if ( n == 0 ) // sets up headers
      {
         for ( int i=0; i < qnumber(header); i++ )
           qstrncpy(arrptr[i], header[i], MAXSTR);
         return;
      }
      // list empty?
      if (!node->altval(NODE_COUNT))
         return;

      node->supval(n-1, &my_movs,sizeof(my_movs));
      outstring = processMemcpy(&my_movs, &movs_distance);
      qsnprintf(arrptr[0], MAXSTR, "%08a", node->altval(n-1));
      qsnprintf(arrptr[1], MAXSTR, "%s", outstring);

      if (movs_distance != BADADDR)
      {
         qsnprintf(arrptr[2], MAXSTR, "%02x", movs_distance);
      }
      else
      {
         qsnprintf(arrptr[2], MAXSTR, "");
      }
      return;
}
/***********************************************************************
* Function: enter
*
* This is a standard callback in the choose2() SDK call. This function
* is called when the user pressed Enter or Double-Clicks on a line in
* the chooser list.
***********************************************************************/
void idaapi enter(void * obj,ulong n)
```

```
{
      ea_t addr;
      netnode *node = (netnode *)obj;
      addr = node->altval(n-1);
      jumpto(addr);
      return;
}
/**********************************************************************
* Function: destroy
*
* This is a standard callback in the choose2() SDK call. This function
* is called when the chooser list is being destroyed. Resource cleanup
* is common in this function. The netnode deleted here.
**********************************************************************/
void idaapi destroy(void* obj)
{
      netnode *node = (netnode *)obj;
      node->kill();
      return;
}
/**********************************************************************
* Function: size
*
* This is a standard callback in the choose2() SDK call. This function
* returns the number of lines to be used in the chooser list.
**********************************************************************/
ulong idaapi size(void* obj)
{
      ulong mysize;
      netnode *node = (netnode *)obj;
      mysize = node->altval(NODE_COUNT);
      return mysize;
}
/**********************************************************************
* Function: functionSearch
*
* functionSearch looks through functions for rep movsd and rep movsb
* memcpy is defined as a rep movsd followed by rep movsb
* single rep movsd and movsb are also recorded
```

```
 *
 * last_movs is used to track for rep movsd/rep movsb sets
 * the netnode's alval and supval arrays are used
 * node->alset contains the base address
 * node->supset contains a movsobj_t object
 *
 *
 * memcpy() == movsobj_t with {addr, addr}
 * mosvd only == movsobj_t with {addr, BADADDR}
 * movsb only == movsobj_t with {BADADDR, addr}
 *
 * NOTE: this function misses rep movw (66 F3 A5) instructions
 ************************************************************************/
void functionSearch(func_t* funcAddr, netnode* node)
{
      movsobj_t my_movs;
      int counter = node->altval(NODE_COUNT);
      ea_t last_movs = BADADDR;
      ea_t addr = funcAddr->startEA;

      while (addr != BADADDR)
      {
         flags_t flags = getFlags(addr);
         if (isHead(flags) && isCode(flags))
         {
            // fill cmd, only looking for 2 byte instructions
            if (ua_ana0(addr) == 2)
            {
               if ((cmd.auxpref & aux_rep) && (cmd.itype == NN_movs))
               {
                 if (cmd.Operands[1].dtyp == dt_dword) // rep movsd
                 {
                    if (last_movs != BADADDR)
                    {
                       // two consecutive rep movsd
                       // set the previous one to movsd only
                       my_movs. movsDW = last_movs;
                       my_movs. movsBT = BADADDR;
                       node->altset(counter, last_movs);
                       node->supset(counter++, &my_movs,sizeof(my_movs));
                    }
```

```
                // found a rep movsd waiting for rep movsb
                last_movs = cmd.ea;
              }
              else if (cmd.Operands[1].dtyp == dt_byte) // rep movsb
              {
                if (last_movs == BADADDR)
                {
                  // rep movsb with no preceding rep movsd
                  my_movs. movsDW = BADADDR;
                  my_movs. movsBT = cmd.ea;
                  node->altset(counter, cmd.ea);
                  node->supset(counter++,&my_movs,sizeof(my_movs));
              }
              else // memcpy()
              {
                  // complete set rep movsd/rep movsb
                  my_movs. movsDW = last_movs;
                  my_movs. movsBT = cmd.ea;
                  node->altset(counter, last_movs);
                  node->supset(counter++, &my_movs,sizeof(my_movs));
              }
              last_movs = BADADDR;
            }
            else
            {
              msg("%x: rep", addr);
              msg("ERROR !!!\n");
            }
          }
        }
      }
      addr = next_head(addr, funcAddr->endEA);
    }

    if (last_movs != BADADDR)
    {
      // a remaining single rep movsd
      my_movs. movsDW = last_movs;
      my_movs. movsBT = BADADDR;
```

```
            node->altset(counter, last_movs);
            node->supset(counter++, &my_movs, sizeof(my_movs));
        }
        node->altset(NODE_COUNT, counter);
        return;
}
/**************************************************************************
* Function: collectData
*
* This function iterates through all functions calling functionSearch
/**************************************************************************
void collectData(netnode* node)
{
        for (uint i = 0; i < get_func_qty(); ++i)
        {
            func_t *f = getn_func(i);
            functionSearch(f, node);
        }
        return;
}
/**************************************************************************
Function: init
*
* init is a plugin_t function. It is executed when the plugin is
* initially loaded by IDA
**************************************************************************/
int init(void)
{
        // plugin only works for x86 executables
        if (ph.id != PLFM_386)
            return PLUGIN_SKIP;
        return PLUGIN_OK;
}
/**************************************************************************
* Function: term
*
* term is a plugin_t function. It is executed when the plugin is
* unloading. Typically cleanup code is executed here.
* The window is closed to remove the choose2() callbacks
**************************************************************************/
```

```
void term(void)
{
      close_chooser(window_title);
      return;
}
/*********************************************************************
* Function: run
*
* run is a plugin_t function. It is executed when the plugin is run.
* This function collects data and and displays results
*
*     arg - defaults to 0. It can be set by a plugins.cfg entry. In this
*           case the arg is used for debugging/development purposes
* ;plugin displayed name    filename       hotkey        arg
* find_memcpy               findMemcpy     Ctrl-F12      0
* find_memcpy_unload        findMemcpy     Shift-F12     415
*
* Thus Shift-F12 runs the plugin with an option that will unload it.
* This allows (edit/recompile/copy) cycles.
*********************************************************************\
void run(int arg)
{
      char node_name[] = "$ inline memcpy";

      if(arg == 415)
      {
         PLUGIN.flags |= PLUGIN_UNL;
         msg("Unloading plugin...\n");
         return;
      }
      netnode* node = new netnode;
      if(close_chooser(window_title))
      {
         //window existed and is now closed
         msg("window existed and is now closed\n");
      }
      if (node->create(node_name) == 0)
      {
         msg("ERROR: creating netnode %\n", node_name);
         return;
      }
```

```
        // set netnode count to 0
        node->altset(NODE_COUNT, 0);

        // look for memcpys
        collectData(node);

        // create chooser list box
        choose2(false,          // non-modal window
           -1, -1, -1, -1,      // position is determined by Windows
           node,                // object to show
           qnumber(header),     // number of columns
           widths,              // widths of columns
           size,                // function that returns number of lines
           description,         // function that generates a line
           window_title,        // window title
           -1,                  // use the default icon for the window
           0,                   // position the cursor on the first line
           NULL,                // "kill" callback
           NULL,                // "new" callback
           NULL,                // "update" callback
           NULL,                // "edit" callback
           enter,               // function to call when the user pressed Enter
           destroy,             // function to call when the window is closed
           NULL,                // use default popup menu items
           NULL);               // use the same icon for all line

        return;
}
char comment[]    = "findMemcpy - finds inline memcpy";
char help[]       = "findMemcpy\n"
                    "This plugin looks through all functions\n"
                    "for inline memcpy\n";
char wanted_name[] = "findMemcpy";
char wanted_hotkey[] = "";

/* defines the plugins interface to IDA */
plugin_t PLUGIN =
{
      IDP_INTERFACE_VERSION,
      0,                  // plugin flags
      init,               // initialize
      term,               // terminate. this pointer may be NULL.
```

```
        run,                // invoke plugin
        comment,            // comment about the plugin
        help,               // multiline help about the plugin
        wanted_name,        // the preferred short name of the plugin
        wanted_hotkey       // the preferred hotkey to run the plugin
};
```

Compile the plug-in and run the plug-in. No longer are we bound to the message window, *findMemcpy* opens a chooser list box similar to Figure 9.17. All the functionality of a built in list box is provided. The list can be sorted by any of the columns. Clicking on a line will jump to the memory address in the disassembly view.

Figure 9.17 Find Memcpy Results

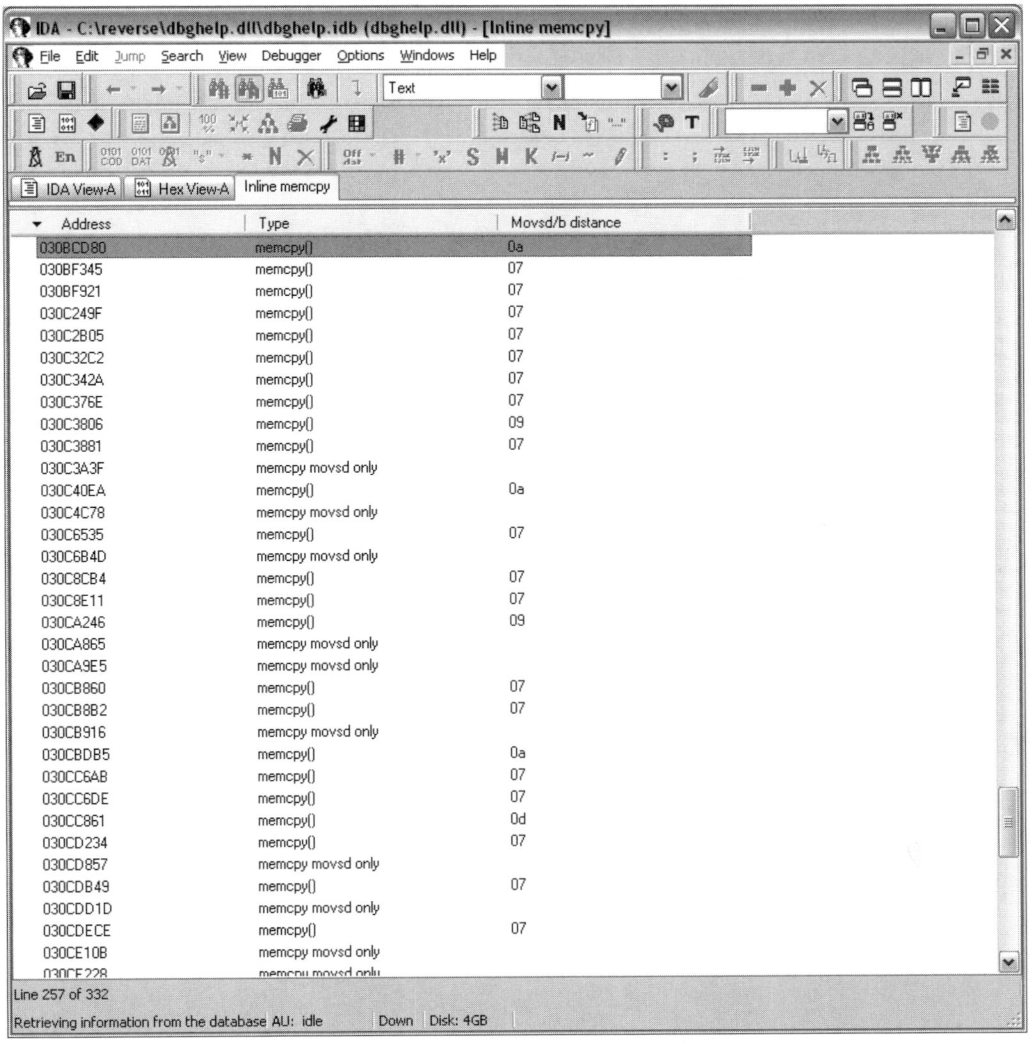

The plug-in introduce some new IDA API functionality, the list box being the most apparent. However the plug-in also introduces one of IDA's built data types. IDA uses the netnode class to store internal information.

What is a netnode? Netnode is defined in *netnode.hpp* and is internally implemented as a B-Tree. Netnode are saved with the database and thus can provide permanent storage tied to an *idb*. This plug-in kills the netnode, since it doesn't require permanence. Two types within netnode are used in both this plug-in and the indirectCall plug-in presented later in the chapter. The types are altval and supval.

Type	Description
altvals	This is a sparse array holding 32 bit values. altvals is often used with addresses as keys. The value bound to the key is then used as an index to the supval array.
supvals	This is an array of variable sized objects (up to MAXSPECSIZE defined as 1024 bytes).

The plug-in uses the supval array to store *movsobj_t* objects. Each *movsobj_t* represents either a *memcpy* or a partial *memcpy*. A partial *memcpy* would be a single rep movsd or *rep movsb*.

```
struct movsObj {
   ea_t movsDW;        // addr of rep movsd. BADADDR if none
   ea_t movsBT;        // addr of rep movsd. BADADDR if none
};
typedef movsObj movsobj_t;
```

If a single or unmatched *rep movsd* is encountered the missing item's address is recorded as *BADADDR*. *Altval* is used as a standard array and the value is the base address to be displayed. The base address is *rep movsd*'s address except for a single *rep movsb*.

Altval holds the array count at index *NODE_COUNT (−1)*. Holding the array count at index −1 is common among other plug-ins as well.

Having covered the data types we can move to the run function. The netnode is created in this function. The use of a '$' prefix to the netnode name is recommended in netnode.hpp. Location names should not be used as IDA names netnodes by location name.

This plug-in uses arguments to the run function. Arguments are defined in the *plugins.cfg* file located in the plugin directory. Add the following to the end of the *plugins.cfg* file:

```
find_memcpy            findMemcpy     Ctrl-F12       0
find_memcpy_unload     findMemcpy     Shift-F12      415
```

Since the plug-in uses callbacks it cannot unload itself after executing run. The extra option is added in order to unload the plug-in. If the proper argument is received the plug-in flags are modified allowing the plug-in to unload. Unloading allows us to compile and copy over a new version of the plug-in without having to shutdown and restart IDA. The section, Plug-in "Debugging Strategies", contains more debugging techniques.

```
if(arg == 415)
{
    PLUGIN.flags |= PLUGIN_UNL;
    msg("Unloading plugin…\n");
    return;
}
```

Collecting Data

The rest of the run calls two functions, *collectData* and the *choose2* API function which creates the list box. The *collectData* function iterates through all the functions calling *functionSearch*, but not before introducing two new API calls. The new functions are defined in *funcs.hpp*.

```
// Get pointer to function structure by number
//      n - number of function, is in range 0..get_func_qty()-1
// Returns ptr to a function or NULL
idaman func_t *ida_export getn_func(size_t n);

// Get total number of functions in the program
idaman size_t ida_export get_func_qty(void);
```

The real work is done by *functionSearch*. The function may look similar to what we saw in IDC. Iterating over areas is very common in scripts and plug-ins. The *while* loop iterates over all the defined items in the function.

```
// Get start of next defined item. Return BADADDR if none exist.
// maxea is not included in the search range

idaman ea_t ida_export next_head(ea_t ea, ea_t maxea);
```

The first *if* statement defines that we are looking for instructions. The next *if* statement includes a new API call.

```
// Analyze the specified address and fill 'cmd'
// This function does not modify the database
// Returns the length of the (possible) instruction or 0

idaman int ida_export ua_ana0(ea_t ea);
```

The *ua_ana0* function part of a family of functions that analyzes instructions defined in *ua.hpp*. The *ua_ana0* is the most minimal as it only analyzes the address without modifying the database. The analysis goes into 'cmd', which holds instruction information.

```
idaman insn_t ida_export_data cmd; // current instruction
```

The *insn_t* type is actually a class which holds both generic and processor specific instruction information. Both *rep movsd* and *rep movsb* are two byte instruction satisfying the *if* statement. The following lines use various *cmd* attributes.

```
01 if ((cmd.auxpref & aux_rep) && (cmd.itype == NN_movs))
02 {
03     if (cmd.Operands[1].dtyp == dt_dword) // rep movsd
04     {
05        // removed for now
06     }
07     else if (cmd.Operands[1].dtyp == dt_byte) // rep movsb
08     {
09        // removed for now
```

Line 1 looks at both the *auxpref* and *itype* attribute for the analyzed instruction. The itype *is* represents the instruction mnemonic.

```
ushort itype;          // instruction code (see ins.hpp)
```

The file *ins.hpp* is not used and instruction nmemonics are located in *allins.hpp*. The instruction nmemonics are stored in very large *enums*. The first two characters define the processor. The *auxref* attribute is a processor dependent field and as *such aux_rep* is bit flag defined in *intel.hpp*. Line 1 thus looks for *rep movsd* or *rep movsb*.

NOTE

What about rep movsw? Intel has these wonderful things called prefixes that can change the sizing of instruction that follows it.

```
   F3 A5     rep movsd
66 F3 A5     rep movsw
   F3 A4     rep movsb
```

The 0x66 prefix is what separates rep movsd from rep movsw. The plug-in will miss rep movsw instructions.

The if statements in lines 3 and 7 look at the operands of the instruction. The *insn_t* class contains an array of *op_t* objects which are the operands.

```
#define UA_MAXOP          6
  op_t Operands[UA_MAXOP];
```

The operand class provides more detail about instructions. Using operands you could determine which register an instruction was using, or if the instruction had an immediate as an offset. The attribute the code uses is *dtype*.

```
#define dt_byte       0       // 8 bit
#define dt_word       1       // 16 bit
#define dt_dword      2       // 32 bit
```

The instruction is now sufficiently decoded to determine *rep movsd* or *rep movsb*. The rest of the code in *funcSearch* looks for matched pairs of *rep movsd/rep movsb*. The variable *last_bp* is used to track the order. Any unmatched moves are also stored. The plug-in does not perform any code analysis and it is possible that there are jumps connecting unmatched sets. The difference between *rep movsd* and *rep movsb* is also calculated in an effort to spot discrepancies.

Displaying Data

IDA has API functions for both single and multi-column list boxes. The *choose2* function is a wrapper that calls choose with preset options such as creating a modal window. The function works well as is relatively easy to use.

The following is the commented prototype from *kernwin.hpp*.

```
inline ulong choose2(
   void *obj,                                  // object to show
   int width,                                  // Max width of lines
   ulong (idaapi*sizer)(void *obj),            // Number of items
   char *(idaapi*getl)(void *obj,              // Description of
   ulong n,char *buf),                         // n-th item (1..n)
                                               // 0-th item if header line
   const char *title,                          // menu title (includes ptr to
                                                  help)
   int icon,                                   // number of the default icon to
                                                  display
   ulong deflt=1,                              // starting item
   chooser_cb_t *del=NULL,                     // cb for "Delete" (may be NULL)
                                               // supports multi-selection
                                                  scenario too
                                               // returns: 1-ok, 0-failed
   void (idaapi*ins)(void *obj)=NULL,          // cb for "New" (may be NULL)
   chooser_cb_t *update=NULL,                  // cb for "Update"(may be NULL)
                                               // update the whole list
                                               // returns the new location of
                                                  item 'n'
```

```
    void (idaapi*edit)(void *obj,ulong n)=NULL,      // cb for "Edit"
                                                     // (may be NULL)
    void (idaapi*enter)(void * obj,ulong n)=NULL,    // cb for non-modal
                                                        "Enter" (may be NULL)
    void (idaapi*destroy)(void *obj)=NULL,           // cb to call when the
                                                     window is closed (may be NULL)
    const char * const *popup_names=NULL,            // Default:
                                                     // insert, delete, edit, refresh
    int (idaapi*get_icon)(void *obj,ulong n)=NULL); // cb for get_icon
                                                     // (may be NULL)
}
```

The following is the call to *choose2* from find *memcpy*.

```
// create chooser list box
choose2(false,            // non-modal window
  -1, -1, -1, -1,         // position is determined by Windows
  node,                   // object to show
  qnumber(header),        // number of columns
  widths,                 // widths of columns
  size,                   // function that returns number of lines
  description,            // function that generates a line
  window_title,           // window title
  -1,                     // use the default icon for the window
  0,                      // position the cursor on the first line
  NULL,                   // "kill" callback
  NULL,                   // "new" callback
  NULL,                   // "update" callback
  NULL,                   // "edit" callback
  enter,                  // function to call when the user pressed Enter
  destroy,                // function to call when the window is closed
  NULL,                   // use default popup menu items
  NULL);                  // use the same icon for all line
```

The find *memcpy* plug-in has many of the callbacks set to *NULL*. However most of the callbacks are not needed. It is not common to add new lines to a list box. The popup menu callback can be useful for operating on list data in ways other than jumping to the disassembly for a single item.

The key callbacks are size, description, enter, and destroy.

The size callback returns the number of lines to display in the list box. There is not much to it unless items are being added or removed from the list. The prototype for find memcpy's size function is:

```
ulong idaapi size(void* obj)
```

The description callback fills in the rows for the list box. It is called for every item in the list. The function is passed the object, line number, and arrptr. The last item is an array of pointers for column data. The description function copies the text data it wishes to display into the array. Description setups the column header when passed 0 for the *n* argument. The following code from find *memcpy* copies the headers column headers into *arrptr.*

```
static const char* header[] = {"Address", "Type", "Movsd/b distance"};
void idaapi description(void *obj,ulong n,char * const *arrptr)
{
    if ( n == 0 ) // sets up headers
    {
      for ( int i=0; i < qnumber(header); i++ )
        qstrncpy(arrptr[i], header[i], MAXSTR);
       return;
}
```

The enter *callback* is generally used to jump to an address. The function is called when the user presses **Ente**r, or double clicks on a line in the chooser list. The function is passed the object and line number.

```
void idaapi enter(void * obj, ulong n)
```

The *destroy* callback is called when the chooser list is being destroyed. Destroy can perform resource cleanup as is the case in find memcpy.

```
void idaapi destroy(void* obj)
{
    netnode *node = (netnode *)obj;
    node->kill();
    return;
}
```

Conclusion

The find *memcpy* plug-in is an introduction to the IDA API. The next plug-in uses and builds upon many of the same functions presented in this section.

The Indirect Call Plug-in

The IDC section presented a script to find and create cross references for indirect calls through a VTable. This solution requires knowing where interesting VTables are located. Instead of observing the targets of a VTable, the opposite approach can be taken by seeking out all the callers. Callers would include any indirect jump instruction. However, for the sake of brevity indirect calls will refer to both indirect calls and jumps.

Proposed Strategy

1. Similar to the find *memcpy* plug-in, the binary is scanned for interesting instructions, in this case, indirect calls.
2. Breakpoints are set on all indirect calls.
3. The plug-in adds a callback to the debugger.
4. The debugger instruments the binary.
5. The callback records information. Optionally the callback performs a *step into* the call target and record the address.
6. Breakpoints are removed when the process exits.
7. Data is presented to the user and optionally cross references are added.

Based on the proposed strategy four separate tasks need to be performed. Separating the tasks allows code to be written for parts that can be replaced at a later point. Fully working chooser lists are not needed immediately, during development writing to the message window will suffice.

- Collect data
- Query user for options
- Implement the callback
- Present results to the user

The plug-in is presented later in the chapter. However, relevant code and screenshots will be shown in the following sections.

Collecting Data

Before starting to collect data, we need data structures to store them in. During the writing of the plug-in the data container changed but netnodes remained the main data structures. The plug-in uses two netnodes and a qvector. Netnodes were introduced in the previous plug-in.

Qvector is also an SDK data type and is defined in pro.h. It supports most of the standard vector methods.

Name	DataType	Description	Internal Type
calls	Netnode	Contains all found indirect calls	indirect_t
vtables	Netnode	Contains all found VTables	vtable_t
bplist	qvector	Index list into calls netnode	ulong

The netnodes use *altval* and *supval* arrays to allow both address lookup as well as iteration of objects. The *altval* sparse array is accessed by address. The value contained in the *atval* is an index into the *supval* array. *Altval* arrays are initialized to 0. Thus supval indexing begins at 1. The following is some example code taken from the *indirectCalls* header comments.

```
// .text:030CC0FB    call   dword ptr [eax+3Ch] ;

indirect_t myObj;

ulong index = calls->altval(0x030CC0FB);

if (index != 0) // indirect call (assume we assigned it earlier)
{
    indirect_t myObj = calls->supval(index, &myObj, sizeof(myObj));
    msg("%a -> %a\n", myObj.caller, myObj.target);
}
```

Collection of data is performed by the *findIndirectCalls* function. The current segment is scanned, not only functions. The function should collect defined indirect calls although not within a proper function. The following code scans for the calls.

```
switch (cmd.itype)
{
case NN_callfi:
case NN_callni:
case NN_jmpfi:
case NN_jmpni:
  {
    if (get_first_fcref_from(cmd.ea) == BADADDR &&
      get_first_dref_from(cmd.ea) == BADADDR) //no fwd xref
```

```
{
  indirect_t currcall;
  fillIndirectObj(currcall);
  if (cmd.itype & NNJMPxI) // jmp?
  {
     currcall.flags |= JMPSETFLAG;
  }
  node->altset(cmd.ea, counter); // altval keyed by addr
  node->supset(counter++, &currcall, sizeof(currcall));
```

The code is similar to the previous plug-in as it analyzes the instruction and checks for certain nmemonics. Cross references checks from the call are performed for both code and data. The data cross reference check is necessary to avoid jump tables. The call to *fillIndirectObj* prepares the *indirect_t* object. Some more instruction decoding takes place which is recorded into the object.

If the call is an indirect near call, further processing takes place. The goal is to determine if the call is of the form:

```
call [reg] or call [reg + offset]
```

The preceding calls particularly with an offset may contain a *VTable* address in the register. The register is extracted and stored in the object along with any offset. Note that the information is located within the operand's type attribute. With the register and offset, the target address can be calculated. In theory all call instructions could be decoded. Finally there is a test checking a flag to determine if a call is a *jmp*. This is done in order to set the appropriate cross reference type if the call is completed during runtime.

This function concludes the collection of information prior to acquiring options from the user.

User Interface

Various options are available to the user. *AskUsingForm_c API* call creates the user interface. (See Figure 9.18)

Certain characters control whether a checkbox or radio button appears. There is not much documentation available; however there is some sample code. (http://www.openrce.org/downloads/details/32/User_Interface_Sample_Code)

Figure 9.18 Indirect Call User Interface

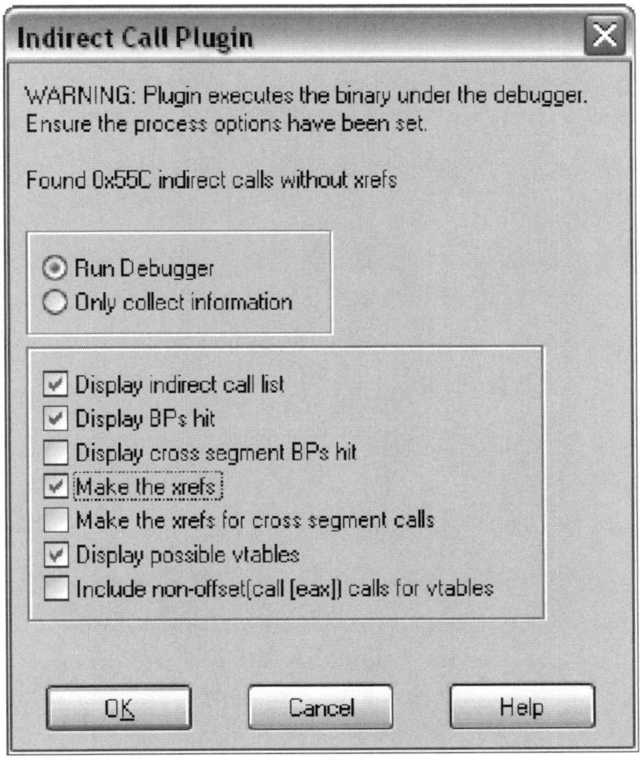

The options are processed and the *if* the user chooses to run the debugger a new API call is made to hook notification of the debugger.

```
if (!hook_to_notification_point(HT_DBG, callback, &gDbgOptions))
{
   warning("Could not hook to notification point\n");
   register_event(E_HOOKFAIL);
   return;
}
```

The following is the function *retype* as well as supported hook types.

```
HT_IDP,          // Hook to the processor module.
HT_UI,           // Hook to the user interface.
HT_DBG,          // Hook to the debugger.
HT_IDB,          // Hook to the database events.
idaman bool ida_export hook_to_notification_point(
                              hook_type_t hook_type,
                              hook_cb_t *cb,
                              void *user_data);
```

The first argument is the type of hook. The second argument is the callback function that receives notification. The final argument can be *NULL*. Passing an object serves two purposes. In order to unhook the same object must be used. The second purpose is passing data to the callback function. In this case the passed *user_data* is a global.

Finally breakpoints are set and the process is started using the *start_process* call.

```
int idaapi start_process(const char *path, const char *args,
                         const char *sdir)
```

If the arguments are *NULL*, *start_process* uses data previously entered under **Debugger | Process options**. The callback is set and the debugger should be running.

Implementing the Callback

The debugger starts and the callback patiently waits for events. The following is the callback's prototype.

```
int idaapi callback(void* user_data,int notification_code,va_list va)
```

The notification code describes the type of event being received. There are various types of notification from low level ones dealing with library loading to higher level breakpoint notifications. The notifications are documented in the *dbg_notification_t* enum located in *dbg.hpp*. The callback has a switch and handles three types of notification.

dbg_bpt

dbg_bpt is the breakpoint notification. The portion of code that handles dbg_bpt has three possible outcomes.

- The breakpoint address is not the calls netnode. This is a user set breakpoint and should be handled as such. The plug-in calls *suspend_process* and exits the callback.

  ```
  suspend_process();
  return 0;
  ```

- The breakpoint was set by the plug-in however the call instruction is not one of the predecoded types. The caller address is stored is *last_bp*, since the target won't be resolved until the *step_into*. A call is made to *request_del_bp and request_step_into*. The *step_into* function cannot be called from a notification handler.

  ```
  last_bp = from;          // saves the caller address
  request_del_bpt(from);   // queue request_del_bpt()
    //
    // From: dbg.hpp request_step_into() AND step into()
  ```

```
  // Type: Asynchronous function - available as Request
  // In Notification handler it is MANDATORY to call
  // Async function in request form
request_step_into();    // queue a request_step_into()
  // request will be run after all notification handlers
run_requests();
break;
```

- The breakpoint was set by the plug-in and the calling instruction is one of the predecoded types. The *my_indirect* object contains both the register number and offset (could be zero). The register is read using *get_reg_val*.

  The register value is then stored into *vtaddr*. This is assumed to be a *VTable* address. In order to recover the target address a *VTable* lookup needs to be performed. However, reading memory while in a notification handler can provide unreliable results. The database and process may not be in sync. The issue was observed during the development of this plug-in. The *invalidate_dbgmem_contents* function invalidates and flushes IDA's cache.

  Inside a notification handler calling *invalidate_dbgmem_contents* is required before reading and writing memory. Another option is *invalidate_dbgmem_config* which although slower is more thorough. Both are defined in bytes.hpp.

  Two more functions are called, *addVTable* and *setTargetXref*. Assuming user options permit, the functions will a create cross reference and possibly a *VTable*.

```
  // copy register_t struct in regval
get_reg_val(regname[my_indirect.call_reg], &regval);
  // vtaddr == VTable base addr
vtaddr = (ea_t)regval.ival;
  // flushes IDA's cache
invalidate_dbgmem_contents((ea_t)regval.ival, 0x100 +
                           my_indirect.offset);
  // read target address from table
to = get_long(my_indirect.offset + vtaddr);
my_indirect.target = to;
  //
addVTable(my_dbg,vtaddr, &my_indirect);
setTargetXref(my_dbg, index, &my_indirect);
calls->supset(index, &my_indirect, sizeof(my_indirect));
del_bpt(from);
continue_process();
break;
```

dbg_step_into

- *dbg_step_into is the step_into* notification. The notification is caused either by the user or the *request_step_into* call. If the user caused the notification, *suspend_process* is called.

  The current *EIP* is the target of the call. The address is copied into the object. *setTargetXref* adds a cross reference based on user options.

```
from = last_bp;
if (from == BADADDR)
{
    suspend_process(); // user caused step_into
    return 0;
}
long index = calls->altval(from); // index into supval
get_reg_val("EIP", &regval); // current EIP is the 'to'
to = (ea_t)regval.ival;
indirect_t my_indirect;
calls->supval(index, &my_indirect, sizeof(my_indirect));
my_indirect.target = to;
    // Add cross reference based on user options and checks
setTargetXref(my_dbg, index, &my_indirect);
    // save completed indirect_t object
calls->supset(index, &my_indirect, sizeof(my_indirect));
    // reset last_bp and continue the debugger
last_bp = BADADDR;
continue_process();
break;
```

dbg_process_exit

This notification signals the termination of the debugged process.

```
unhook_from_notification_point(HT_DBG, callback, user_data);
requestDelBps(calls);
run_requests();
register_event(E_PROCEXIT);

if (options & DISPLAY_INCALLS)
    createIndirectCallWindow(calls);

if (options & DISPLAY_BPS)
    createCompletedBpWindow(calls, my_dbg->bplist);
```

```
    if (options & DISPLAY_VTABLES)
        createVTableWindow(my_dbg->vtables);
```

Presenting Results

The *VTable* display includes an estimated *VTable* size. The size is estimated by iterating through pointers and checking for references. The rest of the presentation functions are similar to the find *memcpy* plug-in. There are two new function introduced in the description callbacks. They both deal with presenting text. The first is *get_nice_colored_name*. This function can construct addresses as seen listed in IDA, such as **segment:address.** Various flags specify the format.

```
#define GNCN_NOSEG      0x0001    // ignore the segment prefix
                                  //producing the name
#define GNCN_NOCOLOR    0x0002    // generate an uncolored name
#define GNCN_NOLABEL    0x0004    // don't generate labels
#define GNCN_NOFUNC     0x0008    // don't generate funcname+… expressions
#define GNCN_SEG_FUNC   0x0010    // generate both segment and function names
     (default is to omit segment name if a function name is present)
#define GNCN_SEGNUM     0x0020    // segment part is displayed as aa hex number
#define GNCN_REQFUNC    0x0040    // return 0 if the address does not
                                  // belong to a function
#define GNCN_REQNAME    0x0080    // return 0 if the address can only be
                                  // represented as a hex number
// returns: the length of the generated name in bytes
// The resulting name will have color escape characters
// GETN_NOCOLOR was not specified
// (see lines.hpp for color definitions)

idaman ssize_t ida_export get_nice_colored_name(
        ea_t ea,
        char *buf,
        size_t bufsize,
        int flags=0);
```

The second new function demangles names. By default IDA uses mangled names, although the option can be changed. This following function produced a short demangled name.

```
inline char *get_short_name(ea_t from, ea_t ea, char *buf, size_t bufsize)
```

Both of the functions are located in *names.hpp*. The plug-in was run against *jscipt.dll* from IE7. Figure 9.19 is the list of all indirect calls.

Figure 9.19 Indirect Call List from jscript.dll

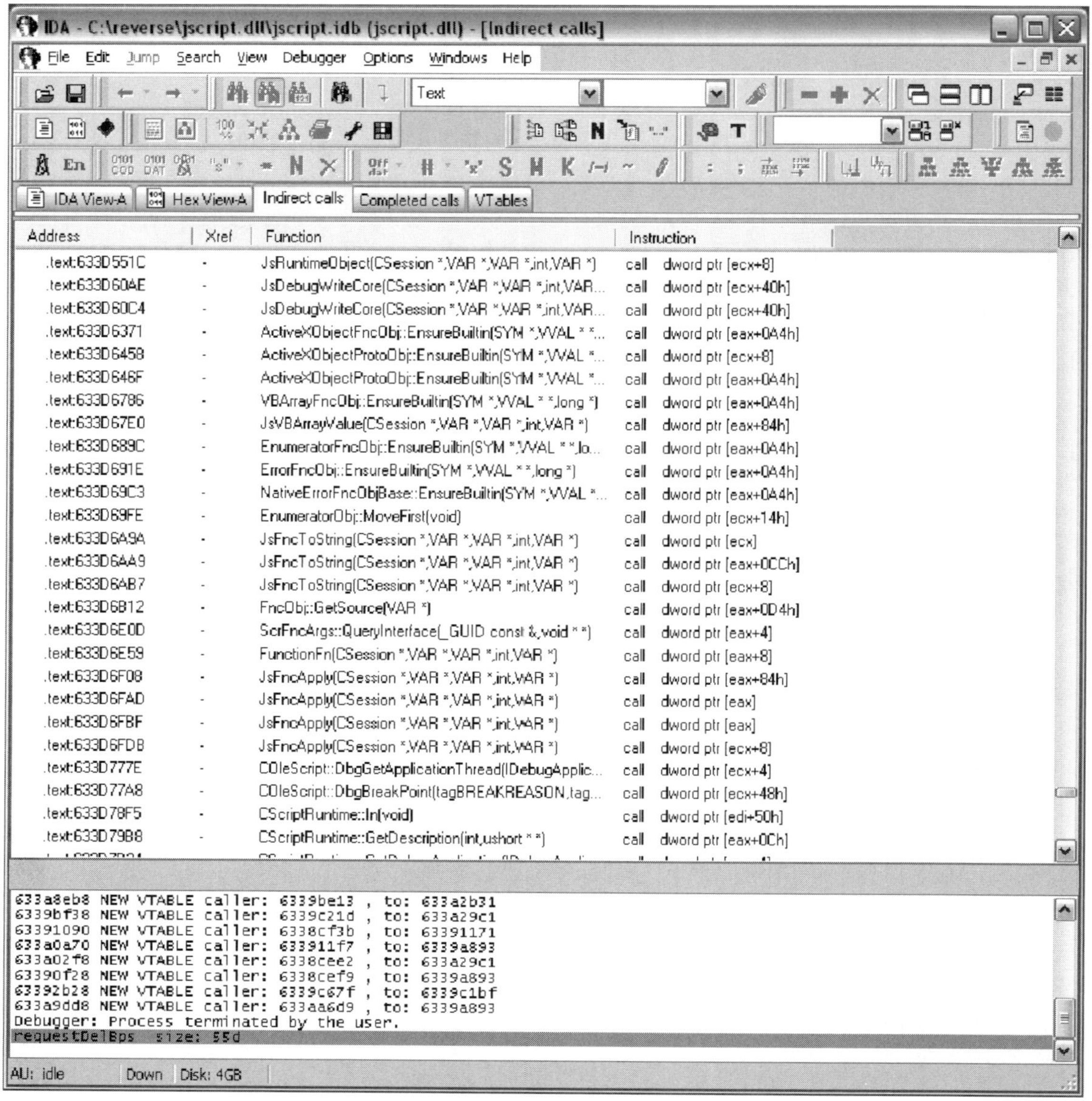

Figure 9.20 Completed Call List from jscript.dll

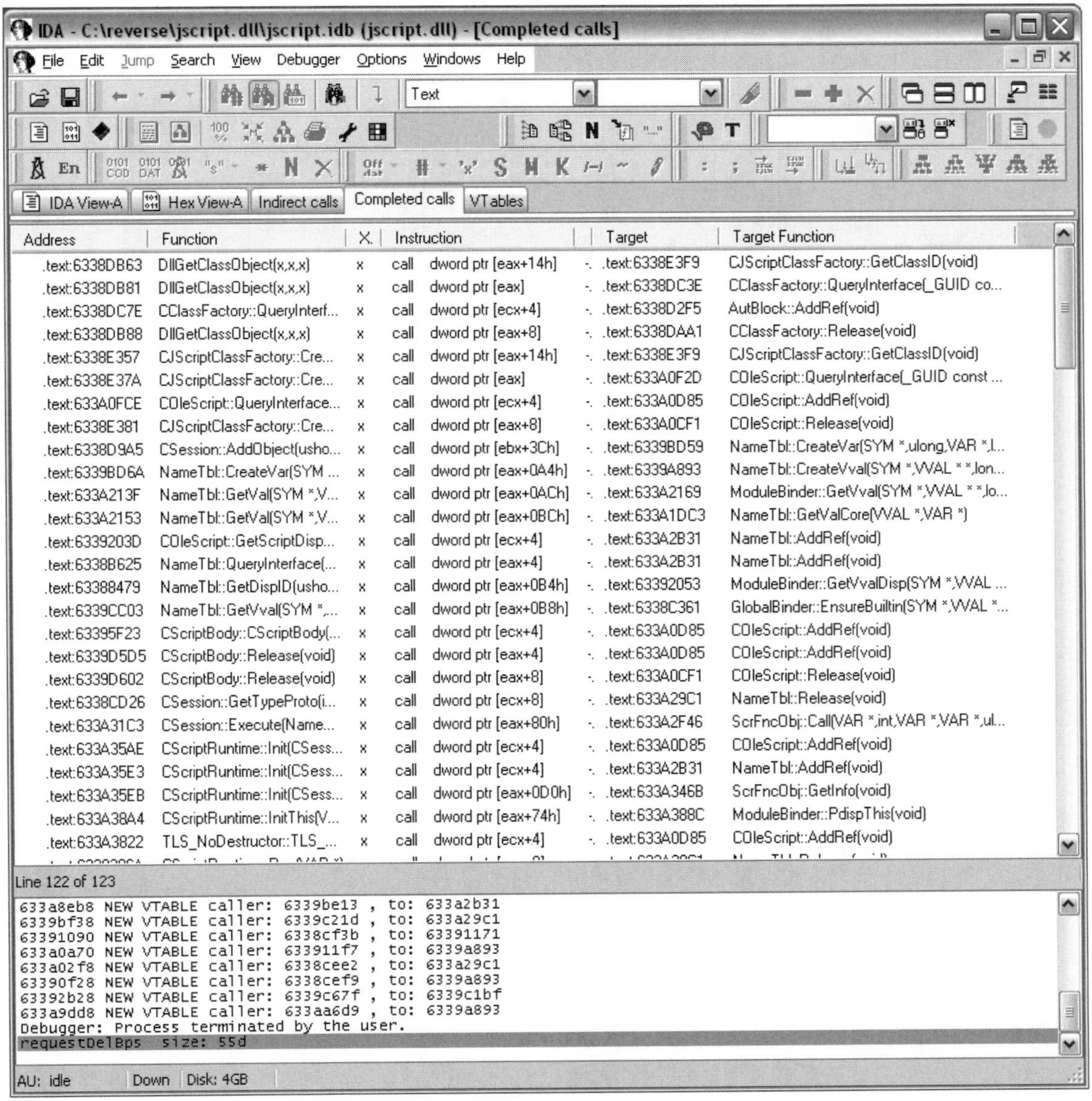

Figure 9.21 VTable List from jscript.dll

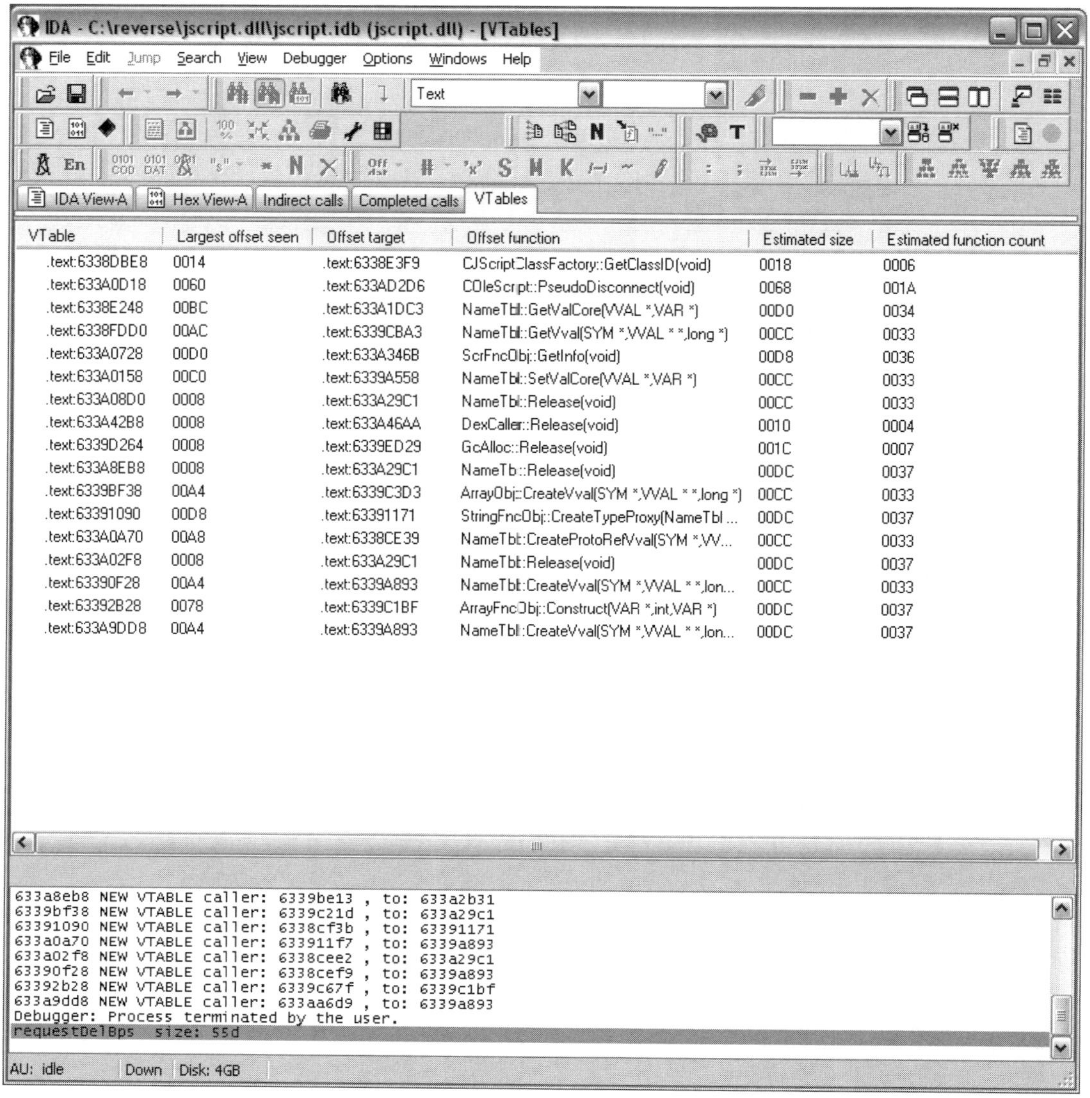

Figure 9.22 Indirect Call Plug-in indirectCall.h

```
/**********************************************************************
* Indirect Call IDA Pro plugin
*
* Copyright (c) 2008 Luis Miras
* Licensed under the BSD License
*
**********************************************************************/
```

```
#ifndef INDIRECTCALLS_H_
#define INDIRECTCALLS_H_
#define NODE_COUNT -1
#define NNJMPxI 0x40
#define CNAMEOPT (GNCN_NOCOLOR | GNCN_NOFUNC | GNCN_NOLABEL)
struct dbgOptions; //fwd declaration
struct indirectCallObj; //fwd declaration
typedef indirectCallObj indirect_t;
typedef qvector<ulong> bphitlist_t;
long vtEstimateSize(ea_t);
void idaapi vtDescription(void *,ulong, char * const *);
void idaapi vtEnter(void * ,ulong);
void idaapi vtDestroy(void*);
void createVTableWindow(netnode* vtables);
void idaapi icDescription(void *,ulong ,char * const *);
void idaapi icEnter(void * ,ulong);
void idaapi icDestroy(void*);
ulong idaapi size(void*);
void createIndirectCallWindow(netnode*);
void idaapi ccDescription(void *,ulong ,char * const *);
void idaapi ccEnter(void* ,ulong);
void idaapi ccDestroy(void*);
ulong idaapi ccSize(void*);
void createCompletedBpWindow(netnode* , bphitlist_t*);
void requestSetBps(netnode*);
void setBps(netnode*);
void requestDelBps(netnode*);
void delBps(netnode*);
void setTargetXref(dbgOptions* , long , indirect_t*);
void addVTable(dbgOptions* , ea_t , indirect_t*);
int idaapi callback(void* , int , va_list);
void fillIndirectObj(indirect_t &);
bool setnodesize(netnode* , long);
long getnodesize(netnode*);
long getobjcount(netnode*);
void findIndirectCalls(segment_t* , netnode*);
void closeListWindows(void);
void register_event(ulong);
void run(int);
```

```
int init(void);
void term(void);
struct indirectCallObj
{
  ea_t caller;     // indirect caller address
  ea_t target;     // target address
  ea_t offset;     // valid for call [reg+offset]
                   // defaults to 0
  short call_reg;  // enum REG
  short flags;     // enum callflags_t
};
struct vtableObj
{
  ea_t baseaddr;       // baseaddr reg in call [reg + off]
  ea_t largestOffset;  // largest off seen in call [reg + off]
};
typedef struct vtableObj vtable_t;
typedef qvector<ulong> bphitlist_t;
struct dbgOptions
{
  netnode* calls;
  netnode* vtables;
  bphitlist_t* bplist;
  ulong options;
};
struct completedbp
{
  netnode* calls;
  bphitlist_t* callindex;
};
typedef completedbp completedbp_t;
enum uioptions_t {
  DISPLAY_INCALLS  = 0x0001,
  DISPLAY_BPS      = 0x0002,
  DISPLAY_XS_BPS   = 0x0004,
  MAKE_XREFS       = 0x0008,
  MAKE_XS_XREFS    = 0x0010,
  DISPLAY_VTABLES  = 0x0020,
  INC_NONOFF_CALLS = 0x0040
```

```
};

enum callflags_t {
  JMPSETFLAG = 1,
  XRSETFLAG = 2,
  XSEGFLAG = 4
};

char* regname[] = {"EAX","ECX","EDX","EBX","ESP","EBP","ESI","EDI"};
enum REG {eax, ecx, edx, ebx, esp, ebp, esi, edi, none = -1};

enum EVENTS
{
  E_START, E_CANCEL, E_OPTIONS, E_HOOKFAIL, E_PROCFAIL,
  E_DWCALL, E_DWXREFS, E_DWVTABLE, E_PROCEXIT
};

// incomplete calls, choose2() list box
char icTitle[] = "Indirect calls" ;
static const char* icHeader[] = {"Address", "Xref","Function", "Instruction"};
static const int icWidths[] = {16, 4, 36, 20};

// completed calls, choose2() list box
char ccTitle[] = "Completed calls" ;
static const char* ccHeader[] = {"Address", "Function", "Xref", "Instruction",
"Xseg","Target", "Target Function"};
static const int ccWidths[] = { 16, 28, 4, 18, 4, 16, 28};

// vtables, choose2() list box
char vtTitle[] = "VTables";
static const char* vtHeader[] = {"VTable ", "Largest offset seen", "Offset
target", "Offset function", "Estimated size", "Estimated function count"};
static const int vtWidths[] = { 16, 16, 16, 28, 16, 20};

// ui string AskUsingForm_c()
const char preformat[] =
"STARTITEM 0\n"
// Help
"HELP\n"
"This plugin searches for indirect calls. For example:\n"
"\n"
"call    dword ptr [eax+14h]\n"
"jmp     eax\n"
"\n"
""
"Breakpoints are set on all the calls.\n"
```

```
"A breakpoint handler will:\n"
" 1. Determine if one of its breakpoints triggered.\n"
" 2. Delete the breakpoint\n"
" 3. Step into the call\n"
" 4. Record both the caller and callee addresses\n"
"\n"
"ENDHELP\n"
// Title
"Indirect Call Plugin\n"
// Dialog Text
"WARNING: Plugin executes the binary under the debugger.\n"
"Ensure the process options have been set.\n\n"
"Found 0x%a indirect calls without xrefs\n\n"
// Radio Buttons
"<#Runs the debugger#"
"Run Debugger:R>\n"
"<#Collects data on indirect calls#"
"Only collect information:R>>\n"
// Check Boxes
"<# Create indirect call window. #"
"Display indirect call list :C>\n"
"<# Create BP window. #"
"Display BPs hit :C>\n"
"<# Include cross segment BPs in BP window. #"
"Display cross segment BPs hit :C>\n"
"<# Automatically create xrefs btwn caller and target. #"
"Make the xrefs :C>\n"
"<# Automatically create xrefs btwn caller and target in different segments. #"
"Make the xrefs for cross segment calls:C>\n"
"<# Create a vtable window #"
"Display possible vtables :C>\n\n"
"<# May lead to false positives (not recommended) #"
"Include non-offset(call [eax]) calls for vtables :C>>\n\n";
#endif /* INDIRECTCALLS_H_ */
```

Figure 9.23 Indirect Call Plugin indirectCall.cpp

```
/************************************************************************
* Indirect Call IDA Pro plugin
*
* Copyright (c) 2008 Luis Miras
* Licensed under the BSD License
*
* Requirements:     This plugin works alongside the IDA Pro debugger.
*                   The plugin requires x86 processor. The plugin "should"
*                   work under the IDA Linux debugger. It has not been
*                   tested.
*
* Description:      The plugin attempt to create cross references for
*                   indirect calls/jmps. For brevity indirect calls/jmp
*                   will be refered only as indirect calls. The plugin
*                   also attempts to identify vtables.
*
* Strategy:         The binary's current segment is scanned for indirect
*                   calls. The binary is instrumented under the debugger.
*                   A breakpoint handler either calculates the target or
*                   steps into the target. Depending on user options cross
*                   references will be made and possible vtables listed.
*
* Data structures:  netnode and qvector are used. Both are built in IDA
*                   types, minimizing 3rd party dependencies. netnodes
*                   allow for persistent data,they are saved in the IDB
*                   However, in this plugin the netnodes are kill()'ed
*
* netnodes are implemented internally as B-trees.
* IDA uses netnodes extensively for its own storage.
* netnodes are defined in netnode.hpp.
*
* netnodes in the plugin: calls - holds all indirect calls
*                         vtable - holds all vtables
*
* netnodes have various internal data structures.
* The plugin uses 2 types of arrays:
*    altval - a sparce array of 32 bit values, initially set to 0.
*    supval - an array of variable sized objects (MAXSPECSIZE)
```

```
*
* Addresses are used as keys into altval array. The value at the key
* is then used as an index into the supval array. The supval array
* holds an object of variable size.
*
* This allows fast lookup using address keys, while being able to
* iterate through all items using supval.
*
* An example:
*
* .text:030CC0FB call dword ptr [eax+3Ch]
*
* indirect_t myObj;
* ulong index = calls->altval(0x030CC0FB);
*
* if (index != 0) // indirect call (assume we assigned it earlier)
* {
*   indirect_t myObj = calls->supval(index, &myObj, sizeof(myObj));
*   msg("%a -> %a\n", myObj.caller, myObj.target);
* }
*
* the calls netnode holds indirect_t objects
* the vtables netnode holds vtable_t objects
* bphitlist_t is a qvector that holds indexes into the calls netnode
*************************************************************************/

#include <ida.hpp>
#include <idp.hpp>
#include <dbg.hpp>
#include <loader.hpp>
#include <allins.hpp>
#include <intel.hpp>
#include "indirectCalls.h"

dbgOptions gDbgOptions = {NULL, NULL, NULL, 0};

/*************************************************************************
* Function: vtEstimateSize
* Args:        ea_t addr       - base address of a VTable
* Return:      long            - Estimated VTable length
```

```
*
* This function attempts to calculate the size of a vtable given its
* base address. It checks xrefs to determine if still in a vtable
*
.text:03010D34 off_3010D34    dd offset sub_308A561
.text:03010D34
.text:03010D38                dd offset sub_3082D8D
.text:03010D3C                dd offset sub_3082DA6
.text:03010D40                dd offset sub_3091542
.text:03010D44                dd offset sub_30B9110
.text:03010D48                dd 75667608h, 6174636Eh, 62h ;
                                                    ; 8 'vfunctab'
.text:03010D54 off_3010D54    dd offset sub_308A561
*
* Sometimes a string is stored at the end of a vtable as in this case.
* vtEstimateSize doesn't understand anything other than dword ptrs
*************************************************************************/
long vtEstimateSize(ea_t addr)
{
  flags_t flags;
  ea_t curraddr = addr;
  ea_t lastaddr = addr;
  bool done = false;
  curraddr = next_head(lastaddr, BADADDR);
  while (!done)
  {
    if (curraddr - lastaddr != 4) // DWORD size differences
    {
      done = true;
    }
    flags = getFlags(curraddr);
    if (!done && !isDwrd(flags))
    {
      done = true;
    }
    // a dref_to could suggest the start of a new vtable
    if (!done && get_first_dref_to(curraddr) != BADADDR)
      done = true;
    if (!done)
```

```
      {
        lastaddr = curraddr;
        curraddr = next_head(lastaddr, BADADDR);
      }
    }
    return lastaddr - addr + 4;
}
/**************************************************************************
* Function: vtDescription
*
* This is a standard callback in the choose2() SDK call. This function
* fills in all column content for a specific line. Headers names are
* set during the first call to this function, when n == 0.
* arrptr is a char* array to the column content for a line.
*                 arrptr[number of columns]
*
* vtDescription creates 6 columns based on the vtHeader array
**************************************************************************/
void idaapi vtDescription(void *obj,ulong n,char * const *arrptr)
{
  netnode *node = (netnode *)obj;
  vtable_t curr_vtable;
  ea_t target;
  long vtSize;
  if ( n == 0 ) // sets up headers
  {
    for ( int i=0; i < qnumber(vtHeader); i++ )
      qstrncpy(arrptr[i], vtHeader[i], MAXSTR);
    return;
  }
  // Empty netnode
  if (!getobjcount(node))
    return;
  char buffer[MAXSTR];
  node->supval(n, &curr_vtable, sizeof(curr_vtable));
  vtSize = vtEstimateSize(curr_vtable.baseaddr);
  target = get_long(curr_vtable.largestOffset + curr_vtable.baseaddr);
  get_nice_colored_name(curr_vtable.baseaddr,
                        arrptr[0], MAXSTR, CNAMEOPT);
```

```
  qsnprintf(arrptr[1], MAXSTR, "%04a", curr_vtable.largestOffset);
  get_nice_colored_name(target, arrptr[2], MAXSTR, CNAMEOPT);

  get_short_name(BADADDR,target , buffer, MAXSTR); //demangles fname
  qsnprintf(arrptr[3], MAXSTR, "%s", buffer);
  qsnprintf(arrptr[4], MAXSTR, "%04a", vtSize);
  qsnprintf(arrptr[5], MAXSTR, "%04a", vtSize/4);
  return;
}
/*************************************************************************
* Function: vtEnter
*
* This is a standard callback in the choose2() SDK call. This function
* is called when the user pressed Enter or Double-Clicks on a line in
* the chooser list.
*************************************************************************/
void idaapi vtEnter(void * obj,ulong n)
{
  vtable_t curr_vtable;
  netnode *node = (netnode *)obj;

  node->supval(n, &curr_vtable, sizeof(curr_vtable));
  jumpto(curr_vtable.baseaddr);
  return;
}
/*************************************************************************
* Function: vtDestroy
*
* This is a standard callback in the choose2() SDK call. This function
* is called when the chooser list is being destroyed. Resource cleanup
* is common in this function. In this case any resource
* cleanup is handled by register_event().
*************************************************************************/
void idaapi vtDestroy(void* obj)
{
  netnode *node = (netnode *)obj;
  msg("\"%s\" window closed\n", vtTitle);
  register_event(E_DWVTABLE);
  return;
}
```

```
/**************************************************************************
* Function: createVTableWindow
*
* A wrapper around choose2() API. 'Generic list chooser (n-column)'
* This sets up the callbacks and necessary options.
* NOTE: 1. Cannot free the "object to show" until chooser closes
*       2. Cannot unload plugin until chooser closes,
*          removing callbacks.
**************************************************************************/
void createVTableWindow(netnode* vtables)
{
  choose2(false,            // non-modal window
    -1, -1, -1, -1,         // position is determined by Windows
    vtables,                // object to show
    qnumber(vtHeader),      // number of columns
    vtWidths,               // widths of columns
    size,                   // function that returns number of lines
    vtDescription,          // function that generates a line
    vtTitle,                // window title
    -1,                     // use the default icon for the window
    0,                      // position the cursor on the first line
    NULL,                   // "kill" callback
    NULL,                   // "new" callback
    NULL,                   // "update" callback
    NULL,                   // "edit" callback
    vtEnter,                // function to call when the user pressed Enter
    vtDestroy,              // function to call when the window is closed
    NULL,                   // use default popup menu items
    NULL);                  // use the same icon for all line
}

/**************************************************************************
* Function: icDescription
*
* This is a standard callback in the choose2() SDK call. This function
* fills in all column content for a specific line. Headers names are
* set during the first call to this function, when n == 0.
* arrptr is a char* array to the column content for a line.
*                 arrptr[number of columns]
*
* vtDescription creates 4 columns based on the icHeader array
```

```
*************************************************************************/
void idaapi icDescription(void *obj,ulong n,char * const *arrptr)
{
  netnode *node = (netnode *)obj;
  indirect_t curr_indirect;

  if ( n == 0 ) // sets up headers
  {
    for ( int i=0; i < qnumber(icHeader); i++ )
      qstrncpy(arrptr[i], icHeader[i], MAXSTR);
    return;
  }

  // list empty?
  if (!getobjcount(node))
    return;

  char buffer[MAXSTR];
  node->supval(n, &curr_indirect, sizeof(curr_indirect));
  func_t* currFunc = get_func(curr_indirect.caller);

  ua_ana0(curr_indirect.caller);

  get_nice_colored_name(curr_indirect.caller,
                       arrptr[0], MAXSTR, CNAMEOPT); // address

  if (curr_indirect.flags & XRSETFLAG)
    qstrncpy(arrptr[1], "x", MAXSTR);
  else
    qstrncpy(arrptr[1], "-", MAXSTR);

  get_short_name(BADADDR, currFunc->startEA, buffer, MAXSTR);
  qsnprintf(arrptr[2], MAXSTR, "%s", buffer);

  generate_disasm_line(cmd.ea, buffer, sizeof(buffer));
  tag_remove(buffer, buffer, sizeof(buffer));
  qsnprintf(arrptr[3], MAXSTR, "%s", buffer);

  return;
}

/*************************************************************************
* Function: icEnter
*
* This is a standard callback in the choose2() SDK call. This function
```

```
* is called when the user pressed Enter or Double-Clicks on a line in
* the chooser list.
*********************************************************************/
void idaapi icEnter(void * obj,ulong n)
{
  indirect_t curr_indirect;
  netnode *node = (netnode *)obj;

  node->supval(n, &curr_indirect, sizeof(curr_indirect));
  jumpto(curr_indirect.caller);
  return;
}

/*********************************************************************
* Function: icDestroy
*
* This is a standard callback in the choose2() SDK call. This function
* is called when the chooser list is being destroyed. Resource cleanup
* is common in this function. In this case any resource cleanup is
* handled by register_event().
*********************************************************************/
void idaapi icDestroy(void* obj)
{
  netnode *node = (netnode *)obj;
  msg("\"%s\" window closed\n", icTitle);
  register_event(E_DWCALL);
  return;
}

/*********************************************************************
* Function: size
*
* This is a standard callback in the choose2() SDK call. This function
* returns the number of lines to be used in the chooser list.
*********************************************************************/
ulong idaapi size(void* obj)
{
  netnode *node = (netnode *)obj;
  return getobjcount(node);
}
```

```
/************************************************************************
* Function: createIndirectCallWindow
*
* A wrapper around choose2() API. 'Generic list chooser (n-column)'
* This sets up the callbacks and necessary options.
* NOTE: 1. Cannot free the "object to show" until chooser closes
*       2. Cannot unload plugin until chooser closes,
*          removing callbacks.
************************************************************************/
void createIndirectCallWindow(netnode* calls)
{
  choose2(false,              // non-modal window
    -1, -1, -1, -1,           // position is determined by Windows
    calls,                    // object to show
    qnumber(icHeader),        // number of columns
    icWidths,                 // widths of columns
    size,                     // function that returns number of lines
    icDescription,            // function that generates a line
    icTitle,                  // window title
    -1,                       // use the default icon for the window
    0,                        // position the cursor on the first line
    NULL,                     // "kill" callback
    NULL,                     // "new" callback
    NULL,                     // "update" callback
    NULL,                     // "edit" callback
    icEnter,                  // function to call when the user pressed Enter
    icDestroy,                // function to call when the window is closed
    NULL,                     // use default popup menu items
    NULL);                    // use the same icon for all line
}

/************************************************************************
* Function: ccDescription
*
* This is a standard callback in the choose2() SDK call. This function
* fills in all column content for a specific line. Headers names are
* set during the first call to this function, when n == 0.
* arg:   arrptr is a char* array to the column content for a line.
*        arrptr[number of columns]
```

```
* arg: completedbp_t* is atruct: netnode*    - points to all calls
*                              bphitlist_t - indexes of hit calls
*
* ccDescription creates 7 columns based on the icHeader array
**********************************************************************/
void idaapi ccDescription(void *obj,ulong n,char * const *arrptr)
{
  completedbp_t* cbp = (completedbp_t*)obj;
  indirect_t curr_indirect;

  if ( n == 0 ) // sets up headers
  {
    for ( int i=0; i < qnumber(ccHeader); i++ )
      qstrncpy(arrptr[i], ccHeader[i], MAXSTR);
    return;
  }

  bphitlist_t& tmp = *(bphitlist_t*)cbp->callindex;
  ulong index = tmp[n-1];

  if (!tmp.size()) // only needed if choose2 kill callback used
    return;        // since it removes members

  char buffer[MAXSTR];

  cbp->calls->supval(index, &curr_indirect, sizeof(curr_indirect));
  func_t* currFunc = get_func(curr_indirect.caller);
  ua_ana0(curr_indirect.caller); //

  // seg.addr
  get_nice_colored_name(curr_indirect.caller, arrptr[0],
                        MAXSTR, CNAMEOPT);

  get_short_name(BADADDR, currFunc->startEA, buffer, MAXSTR);
  qsnprintf(arrptr[1], MAXSTR, "%s", buffer);

  if (curr_indirect.flags & XRSETFLAG)
    qstrncpy(arrptr[2], "x", MAXSTR); // made a cross reference
  else
    qstrncpy(arrptr[2], "-", MAXSTR);

  // get instruction disasm, remove color info
  generate_disasm_line(cmd.ea, buffer, sizeof(buffer));
  tag_remove(buffer, buffer, sizeof(buffer));
  qsnprintf(arrptr[3], MAXSTR, "%s", buffer);
```

```
  if (curr_indirect.flags & XSEGFLAG)
    qstrncpy(arrptr[4], "x", MAXSTR); // cross segment reference
  else
    qstrncpy(arrptr[4], "-", MAXSTR);

  get_nice_colored_name(curr_indirect.target,
                        arrptr[5], MAXSTR, CNAMEOPT);

  currFunc = get_func(curr_indirect.target);
    //demangles fname
  get_short_name(BADADDR, currFunc->startEA, buffer, MAXSTR);
  qsnprintf(arrptr[6], MAXSTR, "%s", buffer);

  return;
}

/************************************************************************
* Function: ccEnter
*
* This is a standard callback in the choose2() SDK call. This function
* is called when the user pressed Enter or Double-Clicks on a line in
* the chooser list.
************************************************************************/
void idaapi ccEnter(void * obj,ulong n)
{
  completedbp_t* cbp = (completedbp_t*)obj;
  bphitlist_t &tmp = *(bphitlist_t*)cbp->callindex;
  indirect_t curr_indirect;
  ulong index = tmp[n-1];

  cbp->calls->supval(index, &curr_indirect, sizeof(curr_indirect));
  jumpto(curr_indirect.caller);
  return;
}

/************************************************************************
* Function: ccDestroy
*
* This is a standard callback in the choose2() SDK call. This function
* is called when the chooser list is being destroyed. Resource cleanup
* is common in this function. In this case any resource cleanup is
* handled by register_event().
************************************************************************/
```

```
void idaapi ccDestroy(void* obj)
{
  completedbp_t* cbp = (completedbp_t*)cbj;
  msg("\"%s\" window closed\n", ccTitle);
  register_event(E_DWXREFS);
  return;
}

/**********************************************************************
* Function: ccSize
*
* This is a standard callback in the choose2() SDK call. This function
* returns the number of lines to be used in the chooser list.
**********************************************************************/
ulong idaapi ccSize(void* obj)
{
  completedbp_t* cbp = (completedbp_t*)obj;
  return cbp->callindex->size();
}

/**********************************************************************
* Function: createCompletedBpWindow
*
* A wrapper around choose2() API. 'Generic list chooser (n-column)'
* This sets up the callbacks and necessary options.
* NOTE: 1. Cannot free the "object to show" until chooser closes
*       2. Cannot unload plugin until chooser closes,
*          removing callbacks.
**********************************************************************/
void createCompletedBpWindow(netnode* calls, bphitlist_t* bplist)
{
  completedbp_t* bp = new completedbp_t;
  bp->calls = calls;
  bp->callindex = bplist;
  choose2(false,             // non-modal window
    -1, -1, -1, -1,          // position is determined by Windows
    bp,                      // object to show
    qnumber(ccHeader),       // number of columns
    ccWidths,                // widths of columns
    ccSize,                  // function that returns number of lines
    ccDescription,           // function that generates a line
```

```
    ccTitle,                // window title
    -1,                     // use the default icon for the window
    0,                      // position the cursor on the first line
    NULL,                   // "kill" callback
    NULL,                   // "new" callback
    NULL,                   // "update" callback
    NULL,                   // "edit" callback
    ccEnter,                // function to call when the user pressed Enter
    ccDestroy,              // function to call when the window is closed
    NULL,                   // use default popup menu items
    NULL);                  // use the same icon for all line
}

/**************************************************************************
* Function: requestSetBps
*
* requests all our breakpoints be set, then run_requests
**************************************************************************/
void requestSetBps(netnode* node)
{
  indirect_t my_indirect;
  long no_calls = getnodesize(node);
  msg("requestSetBps size: %x\n", no_calls);
  for (int i = 1; i < no_calls; ++i)
  {
    node->supval(i, &my_indirect, sizeof(my_indirect));
    request_add_bpt(my_indirect.caller);
  }
  run_requests();
  return;
}

/**************************************************************************
* Function: requestDelBps
*
* requests all our breakpoints be deleted, caller calls run_requests
**************************************************************************/
void requestDelBps(netnode* node)
{
  indirect_t my_indirect;
  long no_calls = getnodesize(node);
```

```
  msg("requestDelBps size: %x\n", no_calls);
  for (int i = 1; i < no_calls; ++i)
  {
    node->supval(i, &my_indirect, sizeof(my_indirect));
    request_del_bpt(my_indirect.caller);
  }
  return;
}

/**************************************************************************
* Function: setBps
*
* set all our breakpoints
**************************************************************************/
void setBps(netnode* node)
{
  indirect_t my_indirect;
  long no_calls = getnodesize(node);
  msg("setBps size: %x\n", no_calls);
  for (int i = 1; i < no_calls; ++i)
  {
    node->supval(i, &my_indirect, sizeof(my_indirect));
    add_bpt(my_indirect.caller);
  }
  return;
}

/**************************************************************************
* Function: delBps
*
* delete all our breakpoints
**************************************************************************/
void delBps(netnode* node)
{
  indirect_t my_indirect;
  long no_calls = getnodesize(node);
  msg("delBps size: %x\n", no_calls);
  for (int i = 1; i < no_calls; ++i)
  {
    node->supval(i, &my_indirect, sizeof(my_indirect));
```

```
    del_bpt(my_indirect.caller);
  }
  return;
}

/*************************************************************************
* Function: setTargetXref
*
* This function serves two purposes. First decides whether to add the
* call to the completed call/bp list. It also can create the cross
* reference between the caller and the target.
*************************************************************************/
void setTargetXref(dbgOptions* myDbg,long index,indirect_t* myIndirect)
{
  bphitlist_t* entry = myDbg->bplist;
  ulong options = myDbg->options;
  ea_t from = myIndirect->caller;
  ea_t to = myIndirect->target;
  short &flags = myIndirect->flags;
  segment_t* from_seg = getseg(from);
  segment_t* to_seg = getseg(to);
  if (from_seg == to_seg)
  {
    if (options & MAKE_XREFS)
    {
      flags |= XRSETFLAG;
      if (flags & JMPSETFLAG)
        add_cref(from, to, (cref_t)(fl_JN | XREF_USER));
      else
        add_cref(from, to, (cref_t)(fl_CN | XREF_USER));
    }
    entry->push_back(index);
  }
  else // cross segment
  {
    if (to_seg != NULL && !(to_seg->is_ephemeral_segm()))
    {
      flags |= XSEGFLAG;
      if (options & MAKE_XS_XREFS)
```

```
        {
          flags |= XRSETFLAG;
          if (flags & JMPSETFLAG)
            add_cref(from, to, (cref_t)(fl_JF | XREF_USER));
          else
            add_cref(from, to, (cref_t)(fl_CF | XREF_USER));
        }
        if(options & DISPLAY_XS_BPS)
        {
          entry->push_back(index);
        }
      }
    }
}

/*********************************************************************
* Function: addVTable
*
* Determines if vtable is considered valid. A new vtable is added to
* the vtable netnode. If the vtable already exists. The offset is
* checked against the largest offset recorded for the vtable.
*********************************************************************/
void addVTable(dbgOptions* myDbg, ea_t vtaddr, indirect_t* myIndirect)
{
  ea_t from = myIndirect->caller;
  ea_t to = myIndirect->target;
  ea_t offset = myIndirect->offset;
  segment_t* from_seg = getseg(from);
  segment_t* vt_seg = getseg(vtaddr);
  netnode* vtables = myDbg->vtables;
  ulong options = myDbg->options;
  if (offset || (options & INC_NONOFF_CALLS))
  {
    if (from_seg != vt_seg) // only documenting vtables in from_seg
    {
      return;
    }
    if ((get_first_dref_to(vtaddr) == BADADDR) ||
      (get_first_dref_from(vtaddr) == BADADDR))
```

```
    {
      msg("%x to %x , probably jump table, not vtable [%x]\n",
           from, to, vtaddr);
    }
    else // considered a valid vtable
    {
      ulong tmp = vtables->altval(vtaddr);
      if (tmp == 0) // new vtable
      {
        vtable_t my_vtable;
        int vtable_counter = getnodesize(vtables);
        my_vtable.baseaddr = vtaddr;
        my_vtable.largestOffset = myIndirect->offset;
        vtables->altset(vtaddr, vtable_counter);
        vtables->supset(vtable_counter++, &my_vtable,
                        sizeof(my_vtable));
        setnodesize(vtables, vtable_counter);
        msg("%x NEW VTABLE caller: %x , to: %x\n", vtaddr, from, to);
      }
      else // vtable already defined
      {
        vtable_t tmpVtable;
        vtables->supval(tmp, &tmpVtable, sizeof(tmpVtable));
        // new offset > old offset
        if (myIndirect->offset > tmpVtable.largestOffset)
        {
          tmpVtable.largestOffset = myIndirect->offset;
          vtables->supset(tmp, &tmpVtable, sizeof(tmpVtable));
        }
      }
    }
  }
}

/************************************************************************
* Function: callback
*
* The debugger calls this function when handling any HT_DBG events.
* The dbgOptions structure is passed to this function allowing the use
* of previously defined data structures and user options.
*
```

```
* callback handles 3 types of HT_DBG events
*
* dbg_bpt - All breakpoints are handled here. The bp address
*           is checked to be ours. If not the the process is
*           suspended. Otherwise:
*           The instruction is call [eax] with or without an
*           offset OR anything else.
*
*           For everything else 'step into' is requested.
*           The current bp addresses is saved in last_bp
*           for the step_into handler
*
*           With the instruction decoded, both the base and
*           target can be calculated.
*           addVTable() & setTargetXref() process if
*           vtables and cross references are made. The
*           indirect_t obj is saved with updates.
*           continue_process() is called
*
*      dbg_step_into - All step_into events are handled here. last_bp
*           is checked. For user caused step_into event
*           suspend_process() is called.
*           setTargetXref() deltemines if cross references
*           are made. The indirect_t obj is saved with updates.
*           continue_process() is called
*
*      dbg_process_exit - This event signifies that the debugger is
*           shutting down. Brealpoints are cleared and depending on
*           options, up to three chooser list windows are opened.
**********************************************************************/
int idaapi callback(void* user_data,int notification_code,va_list va)
{
  dbgOptions* my_dbg = (dbgOptions*)user_data;
  netnode* calls = my_dbg->calls;
  ulong options = my_dbg->options;
  static ea_t last_bp = BADADDR;
  ea_t from = BADADDR;
  ea_t vtaddr = BADADDR;
  ea_t to = BADADDR;
  regval_t regval;
```

```
switch (notification_code)
{
case dbg_bpt:
  {
    va_arg(va, tid_t);
    from = va_arg(va, ea_t);
    long index = calls->altval(from);
    if (index == 0)
    {
      // not one of our breakpoints
      msg("%x not mine options:0x%x", from, options);
      suspend_process();
      return 0;
    }
    indirect_t my_indirect;
    calls->supval(index, &my_indirect, sizeof(my_indirect));
    // check for call [reg] or call [reg + offset]
    if (my_indirect.call_reg == none)
    {
      last_bp = from;
      request_del_bpt(from);
      request_step_into();
      run_requests();
      break;
    }
     get_reg_val(regname[my_indirect.call_reg], &regval);
     vtaddr = (ea_t)regval.ival;
    // flushes possibly stale memory cache
    invalidate_dbgmem_contents((ea_t)regval.ival,
                               0x100 + my_indirect.offset);
    to = get_long(my_indirect.offset + vtaddr);
    my_indirect.target = to;
    addVTable(my_dbg,vtaddr, &my_indirect);
    setTargetXref(my_dbg, index, &my_indirect);
    // save completed indirect
    calls->supset(index, &my_indirect, sizeof(my_indirect));
    del_bpt(from);
```

```
      continue_process();
      break;
    }
  case dbg_step_into:
    {
      from = last_bp;
      if (from == BADADDR)
      {
        msg("not mine\n");
        suspend_process();
        return 0;
      }
      long index = calls->altval(from);
      get_reg_val("EIP", &regval);
      to = (ea_t)regval.ival;
      indirect_t my_indirect;
      calls->supval(index, &my_indirect, sizeof(my_indirect));
      my_indirect.target = to;
      setTargetXref(my_dbg, index, &my_indirect);
      // save completed indirect
      calls->supset(index, &my_indirect, sizeof(my_indirect));
      last_bp = BADADDR;
      continue_process();
      break;
    }
  case dbg_process_exit:
    {
      unhook_from_notification_point(HT_DBG, callback, user_data);
      requestDelBps(calls);
      run_requests();
      register_event(E_PROCEXIT);
      if (options & DISPLAY_INCALLS)
      {
        createIndirectCallWindow(calls);
      }
      if (options & DISPLAY_BPS)
      {
        createCompletedBpWindow(calls, my_dbg->bplist);
      }
```

```
      if (options & DISPLAY_VTABLES)
      {
        createVTableWindow(my_dbg->vtables);
      }
      break;
    }
  default:
    break;
  }
  return 0;
}

/*********************************************************************
* Function: getnodesize
*
* returns size (including location 0)
*********************************************************************/
long getnodesize(netnode* node)
{
  return node->altval(NODE_COUNT);
}

/*********************************************************************
* Function: getobjcount
*
* returns number of items in the netnode not counting invalid slot 0
* see data structure documentation at top of file
*********************************************************************/
long getobjcount(netnode* node)
{
  return node->altval(NODE_COUNT)-1;
}

/*********************************************************************
* Function: setnodesize
*
* store netnode size
*********************************************************************/
bool setnodesize(netnode* node, long size)
{
  return node->altset(NODE_COUNT, size);
}
```

```
/*************************************************************************
* Function: fillIndirectObj
*
* Determines if instruction is call [reg+offset], call [reg], or other
* Fills in the indirect_t struct.
*************************************************************************/
void fillIndirectObj(indirect_t &currcall)
{
  currcall.caller = cmd.ea;
  currcall.target = BADADDR;
  currcall.call_reg = none;
  currcall.offset = 0;
  if (cmd.itype == NN_callni)
  {
    // need a single opcode
    ushort no_operands = 0;
    while(no_operands < UA_MAXOP &&
         cmd.Operands[no_operands].type != o_void)
    {
      no_operands++;
    }
    if (no_operands == 1)
    {
      if (cmd.Operands[0].type == o_phrase)
      {
        currcall.call_reg = cmd.Operands[0].reg;
      }
      else if (cmd.Operands[0].type == o_displ)
      {
        currcall.call_reg = cmd.Operands[0].reg;
        currcall.offset = cmd.Operands[0].addr;
      }
    }
  }
  else if (cmd.itype & NNJMPxI) // jmp?
  {
    currcall.flags |= JMPSETFLAG;
  }
}
```

```
/************************************************************************
* Function: findIndirectCalls
*
* This function through a segment for indirect calls and jmps
* NN_callfi, NN_callni, NN_jmpfi, NN_jmpni
* then it pkgs it in a inidirect_t struct and stores in the netnode
************************************************************************/
void findIndirectCalls(segment_t* seg, netnode* node)
{
  ea_t addr = seg->startEA;
  ulong counter = getnodesize(node);
  while ((addr < seg->endEA) && (addr != BADADDR))
  {
    flags_t flags = getFlags(addr);
    if (isHead(flags) && isCode(flags))
    {
      if (ua_ana0(addr) != 0)
      {
        switch (cmd.itype)
        {
        case NN_callfi:
        case NN_callni:
        case NN_jmpfi:
        case NN_jmpni:
          {
            if (get_first_fcref_from(cmd.ea) == BADADDR &&
              get_first_dref_from(cmd.ea) == BADADDR) //no fwd xref
            {
              indirect_t currcall;
              fillIndirectObj(currcall);
              node->altset(cmd.ea, counter); // altval keyed by addr
              node->supset(counter++, &currcall, sizeof(currcall));
            }
            break;
          }
        default:
          break;
        }
      }
```

```
      }
      addr = next_head(addr, seg->endEA);
    }
    setnodesize(node, counter);
    return;
}
void closeListWindows(void)
{
    close_chooser(icTitle);
    close_chooser(ccTitle);
    close_chooser(vtTitle);
}
/***************************************************************************
* Function: register_event
*
* This function serves as an interface to three semaphores in the form
* of event messages. IDA Pro is single threaded and is non reentrant.
* True concurrency requirements such as mutexes and atomic operations
* are not needed.
*
* The caller reports an event and this function adjusts the semaphores
* and can release resources when needed.
* semaphores are tied to the
*       netnode*      calls   - all indirect calls
*       netnode*      vtables - all vtables
*       bphitlist_t*  bplist  - bp hits, an index list into
*                                         netnode* call
***************************************************************************/
void register_event(ulong rEvent)
{
    static long dbgState = 0;
    static long semcall = 0;
    static long semxref = 0;
    static long semvtable = 0;
    switch (rEvent)
    {
    case E_START:
      {
        closeListWindows();
        if (gDbgOptions.calls)
```

```
        {
          gDbgOptions.calls->kill();
        }
        if (gDbgOptions.vtables)
        {
          gDbgOptions.vtables->kill();
        }
        if (gDbgOptions.bplist)
        {
          gDbgOptions.bplist->~qvector();
        }
        semcall = semxref = dbgState = semvtable = 0;
        break;
      }
    case E_CANCEL:
      {
        semcall = semxref = dbgState = semvtable = 0;
        gDbgOptions.calls->kill();
        gDbgOptions.vtables->kill();
        gDbgOptions.bplist->~qvector();
        break;
      }
    case E_OPTIONS:
      {
        if((~gDbgOptions.options) >> 15)
        {
          dbgState++;
          semcall++;
          semxref++;
          semvtable++;
        }
        if (gDbgOptions.options & DISPLAY_INCALLS)
        {
          semcall++;
        }
        if (((gDbgOptions.options & DISPLAY_BPS) >> 1) && dbgState)
        {
          semcall++;
          semxref++;
        }
```

```
        if (((gDbgOptions.options & DISPLAY_VTABLES) >> 5) && dbgState)
        {
          semvtable++;
        }
        break;
      }
    case E_HOOKFAIL:
      {
        dbgState = semvtable = semxref = 0;
        break;
      }
    case E_PROCFAIL:
      {
        // note: call window may be open
        delBps(gDbgOptions.calls);
        semcall--;
        dbgState = semvtable = semxref = 0;
        unhook_from_notification_point(HT_DBG, callback, &gDbgOptions);
        break;
      }
    case E_DWCALL:
      {
        semcall--;
        if (!semcall)
        {
          gDbgOptions.calls->kill();
        }
        break;
      }
    case E_DWXREFS:
      {
        semxref--;
        semcall--;
        if (!semcall)
        {
          gDbgOptions.calls->kill();
        }
        if(!semxref)
        {
          gDbgOptions.bplist->~qvector();
        }
```

```
      break;
    }
  case E_DWVTABLE:
    {
      semvtable--;
      if (!semvtable)
      {
        gDbgOptions.vtables->kill();
      }
      break;
    }
  case E_PROCEXIT:
    {
      dbgState = 0;
      semcall--;
      semvtable--;
      semxref--;
      if(!semxref)
      {
        gDbgOptions.bplist->~qvector();
      }
      if (!semcall)
      {
        gDbgOptions.calls->kill();
      }
      if (!semvtable)
      {
        gDbgOptions.vtables->kill();
      }
      break;
    }
  default:
    {
      msg("ERROR UNKNOWN EVENT\n");
      msg("%s dbg:%d scall:%d sxref:%d svtable:%d \n",
           "ERROR", dbgState, semcall, semxref, semvtable);
      break;
    }
  }
}
```

```
/***********************************************************************
* Function: run
*
* run is a plugin_t function. It is executed when the plugin is run.
* This function brings up the UI, collects data and sets the debugger
* callback.
*   arg - defaults to 0. It can be set by a plugins.cfg entry. In this
*         case the arg is used for debugging/development purposes
* ;plugin displayed name    filename          hotkey    arg
* indirectCalls_dbg         indirectCalls     Alt-F8    0
* indirectCalls_unload      indirectCalls     Alt-F9    415
*
* Thus Alt-F9 runs the plugin with an option that will unload it.
* This allows (edit/recompile/copy) cycles.
***********************************************************************/
void run(int arg)
{
  char nodename_calls[] = "$ indirect calls";
  char nodename_vtables[] = "$ vtables";
  ea_t curraddr = get_screen_ea();
  segment_t* my_seg = getseg(curraddr);
  char* format;
  short checkbox = DISPLAY_INCALLS | DISPLAY_BPS | DISPLAY_VTABLES;
  short radiobutton = 0;
  int start_status;

  register_event(E_START);

  if(arg == 415)
  {
    PLUGIN.flags |= PLUGIN_UNL;
    msg("Unloading plugin…\n");
    return;
  }

  netnode* calls = new netnode;
  netnode* vtables = new netnode;
  bphitlist_t *hitlist = new bphitlist_t;
  if (calls->create(nodename_calls) == 0)
  {
    calls->kill();
    msg("ERROR: creating netnode %s\n", nodename_calls);
```

```
    return;
  }
  if (vtables->create(nodename_vtables) == 0)
  {
    msg("ERROR: creating netnode %s\n", nodename_vtables);
    calls->kill();
    vtables->kill();
    return;
  }
  calls->altset(NODE_COUNT,1); // position 0 is not used
  vtables->altset(NODE_COUNT,1); // position 0 is not used
  findIndirectCalls(my_seg, calls); // finds jmps/calls
  ulong format_size = sizeof(preformat)+9;
  format = (char*)qalloc(format_size);
  qsnprintf(format, format_size, preformat, getobjcount(calls));
  int ok = AskUsingForm_c(format, &radiobutton, &checkbox); // UI
  gDbgOptions.calls = calls;
  gDbgOptions.vtables = vtables;
  gDbgOptions.bplist = hitlist;
  gDbgOptions.options = checkbox;
  register_event(E_OPTIONS);
  if (!ok)
  {
    msg("user canceled, exiting, unloading\n");
    register_event(E_CANCEL);
    PLUGIN.flags |= PLUGIN_UNL;
    return;
  }
  // debugger closing this window, now only open for non debugger
  if ((checkbox & DISPLAY_INCALLS) && (radiobutton == 1))
  {
    createIndirectCallWindow(calls);
  }
  if (radiobutton == 1)
    return; // only collect data
  // the hook is created here. callback() will receive HT_DBG
  // events only. gDbgOptions is passed to callback()
  // it is global so termination funcs have access
  if (!hook_to_notification_point(HT_DBG, callback, &gDbgOptions))
```

```
  {
    warning("Could not hook to notification point\n");
    register_event(E_HOOKFAIL);
    return;
  }
  requestSetBps(calls);
  start_status = start_process(NULL, NULL, NULL);
  if (start_status == 1) // SUCCESS
  {
    msg("process started…\n");
    return;
  }
  else if (start_status == -1)
  {
    warning("Sorry, could not start the process");
  }
  else
  {
    msg("Process start canceled by user\n");
  }
  register_event(E_PROCFAIL);
  return;
}
/*************************************************************************
* Function: init
*
* init is a plugin_t function. It is executed when the plugin is
* initially loaded by IDA
*************************************************************************/
int init(void)
{
  if (ph.id != PLFM_386) // intel x86
    return PLUGIN_SKIP;
  return PLUGIN_OK;
}
/*************************************************************************
* Function: term
*
* term is a plugin_t function. It is executed when the plugin is
```

```
* unloading. Typically cleanup code is executed here.
* The unhook is called as a safety precaution.
* The windows are closed to remove the choose2() callbacks
*************************************************************************/
void term(void)
{
  unhook_from_notification_point(HT_DBG, callback, &gDbgOptions);
  closeListWindows();
  return;
}
char comment[] = "indirectCalls";
char help[] = "This plugin looks\nfor indirect\ncalls\n";
char wanted_name[] = "indirectCalls";
char wanted_hotkey[] = "Alt-F7";
/* defines the plugins interface to IDA */
plugin_t PLUGIN =
{
  IDP_INTERFACE_VERSION,
  0,                // plugin flags
  init,             // initialize
  term,             // terminate. this pointer may be NULL.
  run,              // invoke plugin
  comment,          // comment about the plugin
  help,             // multiline help about the plugin
  wanted_name,      // the preferred short name of the plugin
  wanted_hotkey     // the preferred hotkey to run the plugin
};
```

Plug-in Development and Debugging Strategies

This section provides some useful strategies to help writing and debugging plug-ins. The Visual Studio debugger works relatively well and is convenient. The debugger can attach and detach to the IDA Process.

Create a new IDA Development Directory

Copy the IDA Pro install directory to a new location, leaving the original directory intact. Choose something short which does not require a lot of typing. For example use:

C:\ida_dev

Go into the plugin directory, in this case **C:\ida_dev\plugins**, and create a new directory called **plugin_backup**. Copy the contents of the plugin directory into the **plugin_backup** directory. Next begin deleting any plug-ins that are not required for the development of the current plug-in. For example if developing a 32 bit plug-in, all the 64 bit plug-ins can be removed. Make sure the keep the debuggers if you are developing a plug-in that uses the debugger.

The removal of the plug-ins serves multiple purposes.

- The message window will contain less extraneous debug messages when using the *–z* debug option, which will be discussed shortly.
- Removing the plug-ins also frees potential hotkeys. We may want to set multiple hotkeys to pass different arguments to the plug-in being developed.
- Startup time decreases without the initialization of unnecessary plug-ins.

Editing Configuration Files

Edit configuration files with testing in mind rather than normal operation. This means removing unnecessary hot key bindings and adding others that may be useful. The following are located in *idagui.cfg*:

Using an Unpacked Database

IDA can operate on the unpacked database which speeds up starting and stopping of the process. When developing a plug-in never use an IDB file without making a backup. In order to operate with unpacked databases do the following:

1. Copy the IDB to a new directory.
2. Make a batch file in the same directory. This file will execute the development copy of IDA. This also allows setting command line arguments. An example *idadev.bat* file could contain the following:

   ```
   C:\ida_dev\idag.exe myidb.idb
   ```

3. Run the batch file. Exit and select **Don't pack database**. You will get a warning, but select **Yes**. The IDB file is no longer in the directory. At this point options can be changed to remove the warnings.
4. Set the following options in *idagui.cfg*:

   ```
   ASK_EXIT_UNPACKED  = NO     // Ask confirmation if the user
                               // wants to exit the database without
                               // packing it
   ASK_EXIT           = NO     // Ask confirmation if the user
                               // wants to exit
   ```

5. Optionally you can assign a hotkey to "Abort".
6. Set the following options in ida.cfg:

```
PACK_DATABASE      = 0      // 0 - don't pack at all
                            // 1 - pack database (store)
                            // 2 - pack database (deflate)
```

The advantages of working with an unpacked database are faster startup and shutdown times.

Note that these options should only be set for the development copy of IDA. The options are not generally recommended.

Enabling Exit without Saving

An alternative strategy is to never save the IDB while developing. The unpacked method will still save the files. If your plug-in crashes the state of the files and database may be unknown. To facilitate operating without saving do the following:

1. Copy the IDB to a new directory.
2. Make a batch file in the same directory. This file will execute the development copy of IDA. This also allows setting command line arguments. An example idadev.bat file could contain the following:

```
C:\ida_dev\idag.exe myidb.idb
```

3. Assign a hotkey to "Abort"

```
"Abort" = "Alt-Z" // Abort IDA, don't save changes
```

When you execute "Abort" a confirmation dialog will come up. There appears to be no options to prevent it, but pressing **Y** will exit.

This strategy has its advantages; the primary one is having a known starting IDB every time in the testing cycle. The downside is somewhat slower start up times. The shutdown time is negligible since IDA doesn't save and pack the database. The amount of delay depends on the size of the IDB.

Plug-in Arguments

Plug-ins can be passed arguments. This can be used to control and change the plug-ins behavior. The *plugis.cfg* file defines hotkeys and arguments to plug-ins.

The IDA API does not allow the unloading of a plug-in. Most non trivial plug-ins will establish callbacks or hooks and remain in memory. This prevents an updated recompiled copy of the plug-in from overwriting the current one. One could exit IDA, but there is a workaround using plug-in arguments. The following is from a *plugins.cfg* file:

```
indirectCalls           indirectCalls     Alt-F8     0
indirectCalls_unload    indirectCalls     Alt-F9     415
```

The corresponding code to handle the argument is:

```
if(arg == 415)
{
  PLUGIN.flags |= PLUGIN_UNL;
  msg(" Unloading plugin…\n");
  return;
}
```

The *PLUGIN_UNL* flag can be set anytime but IDA checks it upon exit of the run function. The plug-in is called with the argument 415, the *PLUGIN_UNL* flag is set. The plug-in should ensure that it unhooks from any notification as well as removing any call-backs. The plug-in using the preceding code performs unhooks in the term function.

Arguments can be used for other things besides unloading the plug-in. An argument could be defined to set a global debug flag. Multiple output functions could exist. For example a certain *arg* could dump results to the message window, while a different arg can create a chooser list box. The argument can be sent from IDC as well, using the *RunPlugin* function.

```
RunPlugin("indirectCalls", 415);
```

Scripting to Help Plug-in Development

Scripting is very useful to test concepts or prototype before writing a plug-in. In particular IDAPython can be very useful since it wraps many of the API calls. IDC can be used as well. Although it lacks some of the more advanced APIs, IDC is always available.

During the development and testing of the indirect calls plug-in, IDC scripts were used. The plug-in uses cross reference data for much of its logic. In order to test and verify that both the plug-in and theories were sound, a script was written. The following is the script.

```
#include <idc.idc>

static decode_xtype(xtype)
{
  if (xtype & XREF_USER)
  {
    Message("XREF_USER");
    xtype = xtype & ~XREF_USER;
  }
  if (xtype == fl_CF)
    Message("fl_CF Call Far");
  else if (xtype == fl_CN)
    Message("fl_CN Call Near");
  else if (xtype == fl_JF)
    Message("fl_JF Jump Far");
```

```
  else if (xtype == fl_JN)
    Message("fl_JN Jump Near");
  else if (xtype == fl_F)
    Message("fl_F Ordinary flow");
  else if (xtype == dr_O)
    Message("dr_O Offset");
  else if (xtype == dr_W)
    Message("dr_W Write");
  else if (xtype == dr_R)
    Message("dr_R Read" );
  else if (xtype == dr_T)
    Message("dr_T Text (names in manual operands)");
  else if (xtype == dr_I)
    Message("dr_I Informational");
}
static lookup_from_ref(void)
{
  auto from, current_code, current_data, no_cxrefs, no_dxrefs;
  from = ScreenEA();
  no_cxrefs = 0;
  no_dxrefs = 0;
  Message("%x [from] xrefs\n", from);
  current_code = Rfirst0(from);
  while(current_code != BADADDR)
  {
    no_cxrefs++;
    Message(" %x CODE (0x%x) ",current_code, XrefType());
    decode_xtype(XrefType());
    Message("\n");
    current_code = Rnext0(from, current_code);
}
  current_data = Dfirst(from);
  while(current_data != BADADDR)
  {
    no_cxrefs++;
    Message(" %x DATA (0x%x) ",current_data, XrefType());
    decode_xtype(XrefType());
    Message("\n");
    current_data = Dnext(from, current_data);
  }
```

```
  if ((no_cxrefs + no_dxrefs) == 0)
    Message(" NONE\n");
}
static lookup_to_ref(void)
{
  auto to, current_code, current_data, no_cxrefs, no_dxrefs;
  to = ScreenEA();
  no_cxrefs = 0;
  no_dxrefs = 0;
  Message("%x [to] xrefs\n", to);
  current_code = RfirstB0(to);
  while(current_code != BADADDR)
  {
    no_cxrefs++;
    Message(" %x CODE (0x%x) ",current_code, XrefType());
    decode_xtype(XrefType());
    Message("\n");
    current_code = RnextB0(to, current_code);
    if (current_code != BADADDR && no_cxrefs > 7)
    {
      Message(" TOO MANY (%d) CODE xrefs …\n", no_cxrefs);
      current_code = BADADDR;
    }
  }
  current_data = DfirstB(to);
  while(current_data != BADADDR)
  {
    no_dxrefs++;
    Message(" %x DATA (0x%x) ",current_data, XrefType());
    decode_xtype(XrefType());
    Message("\n");
    current_data = DnextB(to, current_data);
    if (current_data != BADADDR && no_dxrefs > 7)
    {
      Message(" TOO MANY (%d) DARA xrefs …\n", no_dxrefs);
      current_data = BADADDR;
    }
  }
  if ((no_cxrefs + no_dxrefs) == 0)
```

```
    Message(" NONE\n");
}
static main(void)
{
  AddHotkey("Shift-F7", "lookup_to_ref");
  AddHotkey("Shift-F8", "lookup_from_ref");
}
```

The script binds hotkeys to the lookup functions. Code and data cross references are listed in the message window. While IDA includes *xref.idc*, the format was difficult to read quickly. The following is sample output from *my_xref.idc*, including a listing of the instruction it processed.

```
.text:030BDF13     call     dword ptr [eax+18h] ; my_xref.idc
30bdf13 [from] xrefs
    30cba25 CODE (0x13) fl_JN Jump Near
30bdf13 [to] xrefs
    NONE
```

The script is loaded by *ida.idc*. When a script is included, main does not executed but the functions are available. Thus the hotkeys are bound within *ida.idc*, as shown in the following bit of code.

```
#include <my_xrefs.idc>
static main(void) {
//
//     This function is executed when IDA is started.
//
//     Add statements to fine-tune your IDA here.
//
  AddHotkey("Shift-F7", "lookup_to_ref");
  AddHotkey("Shift-F8", "lookup_from_ref");
}
```

Loaders

Loaders are responsible for recognizing file formats and creating appropriate segments. Analysis is generally performed by processor modules. Loaders as the name implies only load a binary into IDA.

There are various processor modules with source code in the modules directory of the SDK. There are some publically released loaders. *NSDLDR* is a loader for *Nintendo DS ROM* files written by Dennis Elser (http://www.openrce.org/downloads/details/56/NDSLDR). The loader is relatively simple and the code is easy to follow.

Loaders export the *loader_t* structure which is defined in *loader.hpp*.

```
struct loader_t
{
  ulong version;          // api version, should be IDP_INTERFACE_VERSION
  ulong flags;            // loader flags
  accept_file;            // checks the input format. Shows up in the
                          // "load file" dialog box
  load_file;              // loads file into database
  save_file;              // can create output file from database
  move_segm;              // moves segment for relocation or rebasing
  init_loader_options;    // initialize user configurable options
};
```

Processor Modules

Processor modules perform the actual disassembly and analysis of the binary. With over 50 families of processors already supported, most of the major CPUs are covered. However many embedded devices do not have modules yet. Smaller scale devices are built for low cost and such have simpler architectures. These devices can range from standard microcontrollers to rare and limited run chips in audio and video equipment. If there is firmware, someone is going to reverse it.

The SDK has source to many processor modules ranging from the ever popular Atmel AVR chip to the classic z80. Most of the modules use the same file naming convention for each of the main structures allowing for a compare and contrast between modules. The structures used by modules are defined in *idp.hpp*. The main structures are *processor_t*, *asm_t*, and *instruct_t*.

Perhaps writing modules to decode tiny silicon is not to your liking. Modules can and have been written for virtual machines as well. VMs are becoming more popular everywhere from embedded devices to software protections and crackmes. Whether your interest is writing a module for silicon or imaginary silicon, Rolf Rolles' article *Defeating HyperUnpackMe2 With an IDA Processor Module* is a must read, Appendix B in particular. (http://www.openrce.org/articles/full_view/28)

Third-party Scripting Plug-ins

We aren't limited to just writing IDC scripts or full plug-ins in C++. Third party scripting plug-ins provide an alternative. Often using SWIG they wrap many IDC and SDK functions.

The use of a scripting language like Python and Ruby allow access to large libraries of code. Maybe more importantly they bring their nice built in data types. There are currently two choices for scripting languages. Python brought to us in the form of IDAPython and the second is Ruby as IdaRub.

The first scripting plug-in may have been IDAPerl, but it does not appear to be available for download or supported.

IDAPython

IDAPython (http://d-dome.net/idapython) is written by Gergely Erdélyi. It is a very popular plug-in for IDA. New releases focus on coverage of wrapped functions and adding new SDK functions. The source code is available in a Darg repository.

Supported Platforms

IDAPython can run under the Windows or Linux. New test releases are reported to work under Mac OS X.

Installation under Windows is fairly straightforward. IDAPython is available compiled against either Python 2.4 or 2.5. Unless you have specific reasons you want 2.4, install Python 2.5 (http://www.python.org/download/).

Download the appropriate version of the plug-in. I generally use test releases as they will have more wrapped functions. Test releases are hosted here http://code.google.com/p/idapython/.

Unzip the package. Installation consists of copying files to the appropriate places. Copy the python directory to IDA Pro's install directory. Copy *python.plw* from the plug-in directory to IDA Pro's plug-in directory. The plug-in is now installed and will be ready to use the next time IDA is started.

A function reference is available for download or online www.d-dome.net/idapython/reference/. It is generated by *epydoc* directly from the source code.

IDARub

IDARub (http://www.metasploit.com/users/spoonm/idarub/) as implied by the name uses Ruby as its scripting language. IDARub is written by Spoonm. Ruby too has become popular for security tools, the most known being the Metasploit Framework (www.metasploit.com).

The current version of IdaRub is 0.8 was released on August 1, 2006. Since it compiled against the 4.9 SDK, it will continue working with future versions of IDA. However new functions added to the SDK since 4.9 will not be available.

While IDAPython is more popular and supported, there are features only available in IdaRub. Some of the feautures are:

- Remote network access
- Console
- Sweet demos

Sebastian Porst wrote Rublib (http://www.the-interweb.com/serendipity/index.php?/archives/91-RubLib-0.04.html) which is described as a high level API for IdaRub. The current version is 0.04 and it contains over 160 helper functions.

Frequently Asked Questions

Q: Can I make a multithreaded plug-in?

A: IDA Pro is very definitely single threaded. All access to the database would have to be serialized. There are some examples of multithreaded plug-ins. IdaRub written by Spoonm creates a hidden window and handler. Source code is available here: http://www.metasploit.com/users/spoonm/idarub.

Q: My plug-in outputs information to the message window. The message window seems to only hold 2000 lines, can I increase the size of the buffer?

A: IDA can redirect the messages to a log file, if you set the IDALOG environmental variable.. `set IDALOG=mylog.txt`

Q: The list boxes are useful, but can I use the graphing engine for output?

A: The SDK comes with a sample plug-in ugraph which creates a graph view. In the SDK graph.hpp contains the classes relating to graph creation.

Index

Q

R

S